DATE DUE

OC 25 97			
MR 30 '98			
NV 13 '98			
DE 18 '98			
AV 6 99			

HUBBLE

A New Window to the Universe

DANIEL FISCHER   HILMAR DUERBECK

Translated by
Helmut Jenkner and Douglas Duncan

Foreword by Robert Williams

COPERNICUS
AN IMPRINT OF SPRINGER-VERLAG

Originally published as *Hubble: Ein neues Fenster zum All,* © 1995 by Birkäuser Verlag, Basel, Switzerland.

© 1996 Springer-Verlag New York, Inc.

Published in the United States by Copernicus, an imprint of Springer-Verlag New York, Inc.

Copernicus
Springer-Verlag New York, Inc.
175 Fifth Avenue
New York, NY 10010

Manufactured in Italy.

Printed on acid-free paper.

9 8 7 6 5 4 3 2

ISBN 0-387-94672-1 SPIN 10523830

Library of Congress Cataloging-in-Publication Data

Fischer, Daniel.
    [Hubble. English]
      Hubble : a new window to the universe / Daniel Fischer,
Hilmar Duerbeck.
       p.   cm.
      Translated from the German.
      Includes bibliographical references.
      ISBN 0-387-94672-1 (hardcover : alk. paper)
      1. Hubble Space Telescope.   2. Hubble Space Telescope—
History   3. Space Astronomy.   4. Astronomy—Pictorial
works.   I. Duerbeck,
Hilmar W., 1948–   . II. Title
QB500.268.F5813   1996
522′.2919—dc20   96-4818

Contents

Foreword

Astronomy is the oldest of the sciences, and one of the most compelling. Though few people now use it in everyday life, its findings have had a profound influence on how mankind has perceived itself. The discoveries of astronomy have been at the center of some of society's greatest philosophical revolutions. Superficially, astronomy seems to address the nature of "what's out there" as we peer out into space, but fundamentally it addresses the most basic quest of humans: our own origins.

One of astronomy's most amazing discoveries is that most of the atoms in our bodies were created from fiery nuclear reactions deep inside stars. These atoms were blasted into space by gigantic explosions – the death throes of the stars – and new systems like the sun and its planets formed out of this processed material, from which all living things on earth then grew. Thus, when we gaze out on a clear night and observe the seemingly countless specks of light in the sky, we are literally seeing our own origins.

The quest to view the universe with devices such as telescopes, which give a much clearer and brighter picture than the eye, has been one of astronomy's great driving forces. For some centuries this quest led to the construction of progressively larger telescopes, as outlined by Drs. Duerbeck and Fischer in this book, in order to resolve ever finer detail. But in more recent years the earth's atmosphere, which blurs the light we receive from all objects in the heavens, has limited our ability to see the finest detail in astronomical objects. For this reason, ever since the development of large rockets made it a realistic goal, astronomers have wanted to place telescopes in orbit above the earth's atmosphere.

Space astronomy was born after World War II, when sounding rockets carried small telescopes and detectors into space for short periods of time. As payloads gradually became larger and more complex, it became feasible to think of launching a large, multifunctional telescope that could remain in earth orbit and be periodically serviced and refurbished by astronaut crews. The inception of NASA's human space flight program, with its regular shuttle flights, thus made the Hubble Space Telescope project possible.

The Hubble Space Telescope is presently operating in superb fashion every day of the year, acquiring spectacular data from every type of astronomical object in the sky. A variety of instruments and detectors are used to make images and disburse light into spectra, thereby revealing information about the nature of the stars, galaxies, planets, and nebulae.

Technological advances almost always lead to new discoveries. When Galileo, for instance, first turned the newly invented telescope toward the heavens, he found craters on the moon, sunspots, rings around Saturn, and satellites circling the planet Jupiter. The unique capabilities of the Hubble Space Telescope, with its array of instruments, represent a technological advance over previous telescopes that has already produced discoveries such as the strange ways in which dying stars explode and how massive black holes appear to populate the centers of large galaxies.

The beautiful images you see in this book have excited astronomers and the public alike in revealing how intricate the universe is, and they serve as graphic evidence of the complexity of the universe and our own evolution and development.

Robert Williams
Director, Space Telescope Science Institute
May 1995

Preface

Writing a book about the results of five years of Hubble Space Telescope (HST) operations would have taken a similar number of years, had we pursued all of the more than one thousand publications resulting from this project up to the middle of 1995. Hundreds of scientists would have had to be interviewed about their still-unpublished results or new interpretations of earlier observations, and the examination of the extraordinarily colorful history of this billion-dollar enterprise would have had to range from the centers of international space industry to an extended descent into the abyss of American politics. However, all this was not our primary goal. Detailed discussions of individual results we leave to the scientific literature, and the history of the project is presented only in its essentials. With this combination of textual and pictorial information we intend to show that the Hubble Space Telescope is delivering, at least since its repair, spectacular images and data that would be impossible to obtain with ground-based instruments. Furthermore, the knowledge obtained is beginning to change our perception of the universe, even if it may be too early in some cases to discern the nature of these changes. Overall, the investment in a large optical space telescope has paid off in spite of its many troubles. We hope the breathtaking images in this book, together with the sensational results obtained with the Hubble Space Telescope, contribute to convincing the general public of the significance and worth of astrophysical research.

This compilation of Hubble's most interesting results is based mostly, but not exclusively, on publications of the Office of Public Outreach of the Space Telescope Science Institute (ST ScI), which, after a turbulent initial phase during the early 1990s, has reached an exemplary level in trying to disseminate the new results of the Space Telescope to the public. Many of the pictures in this book come from the archive of ST ScI, partly as large format prints, partly as slides, and partly as electronic image files. Special thanks are due to Cheryl Gundy for her help in going through the ST ScI slide collection and to Tim Kimball for a special tour through the institute. The fascinating photos from the space shuttle missions STS-31 and STS-61 for deployment and repair of the Space Telescope were provided by Debbie Dodds of the Johnson Space Center. We acknowledge our helpful discussions with Wolfgang Priester on fundamental questions of cosmology, with Wolfgang Kundt on astrophysics, and with Clark Chapman on planet-related issues (regarding the comet impact on Jupiter).

Historical information about the causes of the Hubble project's initial failures and later (and much greater) successes came from several interviews with witnesses and key participants. Special thanks are due to Riccardo Giacconi and Rudolf Albrecht for several hours of talks; in addition, we would like to thank the latter for a spontaneous invitation to a week-long stay at the Space Telescope European Coordinating Facility in Garching, near Munich, to do research. We were informed about the status of the Hubble project and its future perspectives by Bob Williams, Piero Benvenuti, Richard Hook, Doug Richstone, and Roger Bonnet.

Although cameras represent only a small part of HST's instruments, photographic images and results based on them occupy a major part of this book. For images that are not publicly available we are indebted to Chris Burrows, Rick White, and David Crisp, among others.

Justin Messmer took on the preparation of the materials with his usual enthusiasm. This was an extremely challenging task because of the images' large tonal range. Our editor Thomas Menzel succeeded in combining the parts of the manuscript, which were developed in parallel on different continents, and Georges Batsilas incorporated the corrections

in record time. In addition, we thank Susanne Hüttemeister, Cambridge, MA (who also contributed her knowledge in the area of molecular clouds), and Carlos Hernandez, Miami, FL, for their substantial help in transferring text and images over the Internet, without which this book would hardly have been possible.

Last but not least, we express our cordial thanks to Helmut Jenkner and Doug Duncan, who did a marvelous job of translating, correcting and updating the text, and in revising the third part of the book to make it more informative to readers in the United States.

Daniel Fischer and Hilmar Duerbeck
Königswinter and Santiago de Chile
June 1995

Part 1

From Babylon to Cape Canaveral

Three pioneers of optical astronomy: Galileo Galilei (1564–1642, top left), Isaac Newton (1642–1727, top right), and William Herschel (1738–1822, bottom).

Larger, Higher, More Expensive . . . From Simple Tools to the Telescope of the Nineties

The beginnings of systematic observations of the heavens go back to the third millenium B.C., when Sumerians and Babylonians reflected on the connections between the positions of the stars, the seasons, and other terrestrial events. Everywhere astronomy was needed to measure time, mark the seasons, and guide farmers in their work. In Egypt, the appearance of Sirius marked the beginning of the annual Nile floods. In Central America and China, astronomy and calendrical systems were highly developed.

For thousands of years all observations were done by naked eye, supported by tools for pointing and measuring angles such as the cross staff or the armillary sphere. Not until the first years of the seventeenth century did Galileo Galilei (1564–1642), Thomas Harriot (1560–1621), and others take advantage of the telescope developed by Dutch spectacle manufacturers for terrestrial use, and turn it toward the heavens. These telescopes brought fundamental discoveries that helped astronomy solidify the breakthrough of the Copernican revolution: mountains on the moon, moons orbiting Jupiter, the phases of Venus, resolving the individual stars of the Milky Way.

Isaac Newton (1642–1727) is generally regarded as the founder of modern physical science, not only for his famous laws about the movement of mass and the development of calculus required for their derivation, but also for his theory of diffraction of light. Newton was not only a thinker, however, but a practical man as well. His research into refraction and diffraction of light on glasses, prisms, and mirrors led to the development of the mirror telescope, which he built in 1672 and donated to the Royal Society in London. This invention was critical to the development of astronomy.

What are the advantages of a mirror (reflecting) telescope over a lens (refracting) telescope? A simple lens refracts light of different wavelengths, or colors, differently, so that white light is not collected in a single point; rather, the foci of light of different colors lie at different distances from the lens. Point-like sources of light such as stars therefore appear to be surrounded by a halo of colors. This is called "chromatic aberration." A concave mirror, however, combines light of all wavelengths at the same focus and therefore provides better quality images than a refracting telescope. But soon after Newton's death it was discovered that chromatic aberration could be corrected by combining several lenses – so-called achromats – whereas the manufacturing of well-reflecting mirrors proved to be difficult. The development of mirror telescopes was not pursued further until William Herschel (1738–1822), a musician and amateur astronomer who emigrated from the kingdom of Hanover to England, began to perfect the manufacturing of mirrors toward the end of the eighteenth century. The largest of his mirrors, which were made from metal, had a diameter of 1.2 m.

With these telescopes Herschel revolutionized astronomy. He discovered and catalogued thousands of star clusters and nebulae, and he tried to derive the structure of the Milky Way by star counts. But in spite

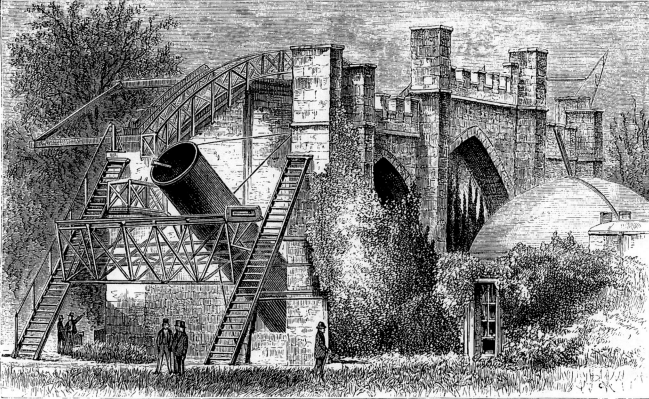

of Herschel's successes – mostly because of the high-quality achromatic objectives of Joseph von Fraunhofer (1787– 1826) – lens telescopes predominated in the large observatories until the end of the 19th century.

Thus Herschel's telescopes remained the largest of their kind for a long time. Only in the middle of the 19th century did the Irish nobleman William Parsons (Earl of Rosse, 1800– 1867) construct a larger one, the so-called Leviathan of Parsonstown, which went into operation in 1845 and had a mirror diameter of 1.8 m. But technical, climatic, and social problems rendered it almost useless. The difficulty of handling it, the foggy Irish climate, and the famines of that time, which drove many people into emigration, diverted the Earl's interests toward different issues. One fundamental discovery was made with this telescope, however: the observation that some nebulae that could not be resolved into stars had a spiral structure. The spiral nebulae became a separate class of celestial objects, but their true nature was not revealed until 75 years later.

At about the same time the director of the Edinburgh Observatory, Charles Piazzi Smyth, looking for a better observing site, investigated the climate of the Canary Islands. In 1856 he conducted a first expedition to Tenerife, maintained a temporary observing site for several weeks, and discovered that the steadiness and clarity of the air on this mountainous island were much better than at lower altitudes – a discovery that would prove to be of greatest importance to astronomy.

Piazzi Smyth's idea of building observatories on mountaintops fell on fertile ground, particularly in the United States. The planning and construction of the Lick Observatory in the vicinity of San Francisco make a good example of how a major observatory is developed. First, a sponsor was required who would provide the funds for such a project. He was found in the person of the former carpenter James Lick, who profited from the great gold rush and his piano manufacturing business to become one of the richest men of California. At first, he intended to have his tomb built in the form of a pyramid in the center of San Francisco, but then decided to provide $700,000 (about $10 million in current dollars) for an observatory on the summit of Mount Hamilton, at an altitude of 1,200 m, near San José, California. Af-

Lord Rosse's telescope, the famous Leviathan, at Birr Castle, Ireland (Source: Simon Newcomb: Popular Astronomy, New York, Harper & Brothers, 1878).

The large refractor at Lick Observatory – with a lens diameter of 0.91 m, one of the largest refracting telescopes in the world (Source: Astronomical Society of the Pacific).

ter Lick died, in 1876, his casket was incorporated in the observatory's foundation. Like most large observatories of the 19th century, the Lick was equipped with a refracting telescope – at the time of its installation the largest in the world, with an objective lens diameter of 0.91 m and a focal length of 17.6 m. It was mainly used for visual observations of binary stars, to discover the fifth moon of Jupiter, and also for photographic images of the Moon. As a second telescope, a "used" 0.93 m reflector by the English amateur Edward Crossley was installed, but it was hardly used at first because of numerous technical problems. But the astronomers wanted more. While Edward E. Barnard, a pioneer of celestial photography, made pictures of bright clouds in the Milky Way and of dark nebulae using portrait lenses, James Keeler totally refurbished the Crossley Telescope and in 1898 began to use it to photograph gaseous and spiral nebulae. In 1904, the Crossley reflector received a new mounting (an English mounting) and helped establish the importance of photography in astronomy. In this form the telescope remains in service today.

Another major American observatory, the Yerkes Observatory, was funded by the somewhat dubious streetcar and realty magnate Charles Yerkes and built under the leadership of George Ellery Hale (1868–1938) in 1897. It was equipped in similar fashion to the Lick Observatory – but with an even larger refractor (1.02 m diameter, 19.4 m focal length). The Yerkes Observatory is situated a few hours' drive from the University of Chicago, only a little above sea level. Aside from the refractor (the largest that has ever been built), the observatory received a 0.60-m reflecting telescope constructed by George W. Ritchey. But Hale was not content with this. Drawing on his own funds and on a large contribution from the steel magnate Andrew Carnegie, in 1908 he built Mount Wilson Observatory on top of Mount Wilson, at an altitude of 1,800 m, in the vicinity of Los Angeles, California. Aside from several telescopes for solar observations, Mount Wilson Observatory has two large reflectors as main instruments, a 1.5-m and a 2.54-m telescope; the latter, the largest telescope of its time, was used by Edwin Hubble for his famous extragalactic investigations during the 1920s, which demonstrated the expansion of the universe (see pp. 16–17).

The course of optical astronomy in the twentieth century had been set: higher, larger, more expensive. Let us look at a number of examples on this path.

Why is the Space Telescope called Hubble?

Naming satellites for important scholars and scientists is a common practice. For instance, the ultraviolet satellite OAO-3, launched in 1972, was called Copernicus; a Jupiter probe launched in 1989 was called Galileo, and the X-ray satellite OAO-4 Einstein. This last name was given not only to honor the father of relativity theory, but because the satellite was expected to observe many objects containing neutron stars and black holes – objects predicted by the theory of relativity. In similar fashion, two reasons influenced the naming of the Space Telescope. First, one of the prime areas of research with the telescope would be the determination of the expansion rate and the age of the universe – the main areas of interest of the American astronomer Edwin Hubble. Second, the Space Telescope was named after Hubble in deference to one of the greatest astronomers of the twentieth century, a discoverer of the Big Bang.

Who was Edwin Powell Hubble (1889–1953)? He spent his childhood in Kentucky and then moved to Chicago, where his father, a lawyer, worked for an insurance company. He was a gifted pupil and a good athlete, who once considered a career as a professional boxer and would later work

Edwin Powell Hubble (1889–1953).

briefly as a basketball coach. He received a fellowship to study at the University of Chicago and later at Oxford, England. However, in England he studied law, and practiced for a few years. Because this did not satisfy him, he went back to the University of Chicago, and worked at the Yerkes Observatory with Prof. F. B. Frost. In 1917 he completed his doctoral thesis on gas nebulae. The director of Mount Wilson Observatory, George Ellery Hale, offered him a position, which he accepted two years later, after he did volunteer duty in Europe during World War I from 1917 through 1919.

With the newly commissioned 2.54-m telescope of the Mount Wilson Observatory, Hubble investigated the Andromeda galaxy (M31) and its variable stars. First, he only detected novae, as others before him had done in M31 and other galaxies. He finally found regularly pulsating stars in 1923. Those were so-called Cepheids, which had been discovered previously in the Magellanic Clouds by Henrietta Leavitt, and for which Ejnar Hertzsprung, Harlow Shapley, and others had established a period-luminosity relation. If the period of one of these stars is obtained, its luminosity is known as well. Comparing its luminosity with the apparent magnitude, its distance can be determined. Hubble was able to derive the distance to the Andromeda galaxy in this way. Although his value of 800,000 light years was too low, the order of magnitude was correct. He had established once and for all that spiral and elliptical galaxies were objects outside our own Milky Way.

In 1929 Hubble analyzed the distances of galaxies, for which redshifts in their spectra had been measured by Vesto Slipher years earlier. First, he restricted himself to galaxies with distances less than 6 million light years, and found a linear relationship between the redshift and the distance. During the following years, supported by the observer Milton Humason, he extended this investigation to distances of up to 100 million light years – and the linear relationship manifested itself even more clearly. This relationship yields the so-called Hubble law: The more distant a galaxy, the larger its redshift. Hubble had thus found several items of importance for the future of astronomy. The Milky Way is not the only galaxy in the universe; rather,

there are infinitely many, receding from each other. Therefore the universe expands, and so must have originated at a single point. This realization fit well with the expanding models of the universe based on the general theory of relativity that had been developed by Willem de Sitter (1917), Alexander Friedman (1922), and Georges Lemaitre (1927). Hubble proved by his observations with what was then the largest telescope in the world, and by his empirically derived law of the expansion of the universe, that our universe was consistent with the theoretically derived Friedmann-Lemaitre model, which postulates an origin in the Big Bang and expansion for billions of years – maybe forever. Hubble's investigations of the spatial distribution of galaxies led him to believe that their distribution was uniform over large scales. His book "The Realm of the Nebulae" is worth reading even today, as it provides an excellent overview of extragalactic research during the first third of the twentieth century.

World War II interrupted Hubble's astronomical studies. He worked in a ballistic research lab for military applications. After the war, he had but little opportunity to use the new telescopes of the Mount Palomar Observatory. He died, unexpectedly, of a stroke in 1953.

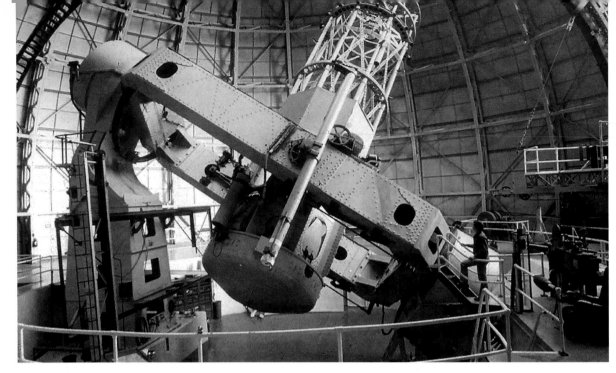

As the successor, as it were, of the Mount Wilson Observatory, the Palomar Observatory (1,700 m altitude) contained the world's largest telescope for many years: a Cassegrain telescope with a mirror 5.08 m in diameter, which was cast in 1934. The telescope itself became operational in 1948. Edwin Hubble, the master himself, inaugurated the "Big Eye."

During the 1970s, a number of large reflectors were put into service. Telescopes with 4.0 m diameter mirrors were installed at Kitt Peak National Observatory in the United States, at Cerro Tololo Interamerican Observatory in Chile, and at the Anglo-Australian Observatory in Australia. Telescopes with 3.6 m diameter mirrors went into operation at the European Southern Observatory in Chile and in the Canada-France-Hawaii Telescope in Hawaii. After these telescopes, which adhered to the "classical" principles of construction, new designs were carried out utilizing computer-aided pointing and mirror adjustment: the first large optical telescope with alt-azimuth mounting (mirror diameter of 6.0 m) of the Special Astrophysical Observatory Zelenchukskaya in the Caucasus Mountains (operational since 1976); the Multiple Mirror Telescope on Mount Hopkins in Arizona, consisting of six 1.82-m telescopes on a single alt-azimuth mounting (with an effective mirror diameter of 4.5m); and the Keck telescope (Mauna Kea, Hawaii, operational since 1992), with a main mirror of 10.0-m diameter, consisting of 38 segments. The largest telescope ever planned, and currently under construction, is the Very Large Telescope (VLT) of the European Southern Observatory (ESO) at Cerro Paranal in northern Chile. This consists of four 8-m telescopes, the first of which will become operational around the turn of the century. All of these telescopes of the 1990s feature thin, light, elastic mirrors whose deformations are compensated by computer-controlled mirror support systems. These "active optics" maintain the optimal shape of the mirror surface under all positional and temperature conditions. Another step in improving image quality is "adaptive optics," in which the wavefront distortions caused by the Earth's atmosphere are corrected. First successes using this technique have been achieved at long wavelengths. The development of further image improvement techniques on large ground-based telescopes opens the door to fascinating possibilities for optical astronomy from the Earth's surface. One day, images from ground-based telescopes may reach the quality of those from the Hubble Space Telescope. But for now the Hubble is unrivaled – and in the ultraviolet region of the spectrum it will always remain so because of the opaqueness of the atmosphere.

The 2.54 m Hooker reflector at the Mount Wilson Observatory near Los Angeles – the largest telescope in the world during the 1920s. Many of Edwin Hubble's observations were made with this telescope (Source: ASP).

Space Astronomy – Picture Window to the Universe

By building bigger and better telescopes we can use light more and more efficiently as a tool for exploring the universe. But what is light? Are there other kinds of radiation besides light, or any other tools to probe the secrets of the universe?

Light is an electromagnetic wave, a form of radiation, that propagates through space with a speed of 300,000 kilometers per second. It is characterized by its wavelength, and our eye registers light of different wavelengths as different colors. A prism, for instance, disperses white light, which consists of light of different wavelengths, into its component colors – the colors of the rainbow. Around 1800 it was recognized that in addition to visible light of different colors there must exist different kinds of radiation. A thermometer positioned beyond the red light behind a prism indicates such a radiation: the longer-wavelength infrared radiation sometimes called thermal radiation; we sense it as heat. Beyond the violet light exists a kind of radiation that is capable of exposing photographic paper: the shorter-wavelength ultraviolet radiation. Heinrich Hertz (1857–1894) succeeded in generating radiation with even longer wavelengths – radio waves. Between radio and infrared are microwaves, which we use for cooking. And Wilhelm Conrad Röntgen (1845–1923) discovered the extremely short-wavelength X-rays. It was to take several decades before astronomers were able to explore these areas of the electromagnetic spectrum.

The various kinds of radiation can be characterized by their wavelengths, their frequencies, or by the energy of their photons. The shorter the wavelength of light, the more energy its photons carry. Most wavelengths cannot penetrate the Earth's atmosphere, and astronomers speak of "windows" for wavelengths that do. The "radio window," the area of the long-wavelength (low-frequency) radio waves, ranges from a wavelength of several dozen meters to 1 cm. It is followed by microwaves and the infrared (several millimeters to 1/10,000 cm), for which the Earth's atmosphere has little or no transmissivity. Next comes the narrow optical window of visible light (radiation with wavelengths from 700 to 300 nm). From 300 nm to 10 nm is the ultraviolet part of the spectrum, and then the area of X-rays. Extremely short-wave radiation, with less than 0.001 nm, is called gamma radiation.

All of this short-wavelength, high-energy radiation is absorbed by the Earth's atmosphere, which is rather important for the well-being of life on Earth – a fact that can be confirmed by anybody who has suffered from sunburn due to excessive exposure to ultraviolet radiation. Only visible light, narrow ranges of infrared, and radio waves can penetrate the Earth's atmosphere, and astronomers can explore the universe from the ground only through these windows. For obvious reasons, astronomy has concentrated on these wavelength ranges, first the visible light, and then, in the first half of the 20th century, the cosmic radio waves. The latter, however, were discovered not by an astronomer but by a radio technician. After attempts to observe the solar radiation in the radio range had failed around the turn of the century, astronomers lost interest in radio astronomy. But Karl Jansky (1905–1950) discovered sources of radio emission in the sky, and his investigations were continued by Grote Reber (b. 1911) during the 1940s. Only after the end of World War II, as radar technology began to look for new applications, did radio astronomy's glory days begin. As with large optical telescopes at the end of the 19th century, the following decades saw the construction of more and more powerful radio telescopes for exploring long-wavelength radiation from space. The most powerful radio telescope today is the Very Large Array (VLA) in the plains of San Augustin, New Mexico, consisting of 27 identical parabolic antennae with a diameter of 24.9 m each, distributed over an area of several square kilometers. This is a telescope that takes half an hour to drive around in a car!

But what happened to the other parts of the electromagnetic spectrum? To observe the other ranges of radiation from space, which are being absorbed fully or at least in large part by the Earth's atmosphere, we had to conquer space itself. Thus, with the advent of the age of space flight, the age of space astronomy began as well. The beginnings were very modest. It did not start with the construction of pure research satellites. Rather, space astronomy turned out to be a child of war, and it was carried out as a by-product of military missions. Some of the confiscated ballistic V2 rockets of the German army were fitted with research equipment in the United States and launched. These experiments yielded the first ultraviolet spectra of the sun. An experiment to measure X-rays originating from the moon led to the detection of X-ray stars; the Italian scientist responsible for this, Riccardo Giacconi, will continue to appear in this book. Satellites with gamma ray detectors, looking for nuclear explosions as part of the verification of the nuclear weapons test ban, discovered strange flashes of gamma radiation of cosmic origin. Progress was made in the long-wavelength range of the spectrum as well. Since infrared radiation is absorbed mainly by water vapor in the atmosphere, there are infrared telescopes on high mountaintops, in high flying aircraft and balloons, and in space.

The first successful ultraviolet observing satellite was the OAO-2. It was launched on December 7, 1968, with a payload of several photometers, which measure the brightness of incoming radiation, and spectrographs, which separate the radiation into different wavelengths. This satellite was developed by the University of Wisconsin and the Smithsonian Astrophysical Observatory at Cambridge, Massachusetts. It was successful in investigating the attenuation of radiation as it traverses the cosmic dust and in the discovery of galaxies and stars with an excess of ultraviolet radiation. On August 21, 1972, OAO-3 (Copernicus) was launched, carrying a 0.80-m telescope with a high-resolution spectroscope and various smaller X-ray telescopes. Its main area of exploration was also interstellar matter. Evidence for the

Three famous astronomy satellites: the International Ultraviolet Explorer (IUE) of NASA and ESA (top left), the German X-ray satellite ROSAT (bottom), and the Compton Gamma Ray Observatory (GRO) of NASA (top right), at 17 metric tons the heaviest research satellite in the history of space flight (Sources: ASP and NASA).

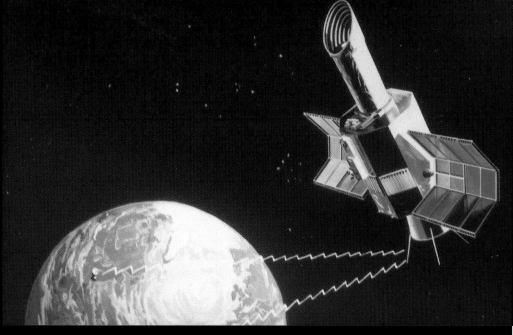

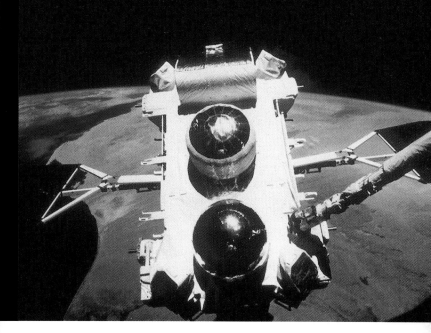

existence of a hot gas in the near reaches of the universe, into which clouds of cool gas are embedded, is due to data from this satellite.

Europe played an important role in ultraviolet astronomy as well. The TD-1 satellite of ESRO (the European Space Research Organization), one of the two precursors of the European Space Agency, was launched on March 12, 1972, and remained operational for two years. Its main mirror of 27.5 cm deflected the ultraviolet radiation to a spectrometer operating in the range from 135 to 255 nm. Another spectrometer covered three areas between 206 and 287 nm. This satellite resulted in an atlas of ultraviolet spectroscopy of bright stars. Another satellite, the Astronomical Netherlands Satellite, was in operation from 1974 to 1976. It was used to perform ultraviolet photometry for a large number of stars. The most successful ultraviolet satellite is the International Ultraviolet Explorer (IUE), built in Great Britain and launched into geosynchronous orbit on January 26, 1978. It consists of a 0.4-m telescope and spectrographs for high and low resolution in the range from 115 to 320 nm. Using two ground stations, astronomers can obtain data on bright and faint planets, stars, nebulae, and galaxies, for 24 hours per day under real-time control. On December 31, 1997, after almost 20 years of service, this satellite will probably be switched off for good. Other satellites, such as the Extreme Ultraviolet Explorer, are geared toward far-ultraviolet observations.

Satellites for X-ray astronomy are even more numerous. The first probe to perform an X-ray survey of the sky was Uhuru, launched in 1970. It discovered 339 sources, including X-ray binaries, supernova remnants, Seyfert galaxies, and clusters of galaxies. Other satellites that were deployed to analyze specific objects include the U.S. Einstein satellite (1978–1981) and the ESA probe EXOSAT (1983–1986). A new sky survey was carried out in the 1990s by the German ROSAT. A number of Japanese satellites (Ginga, Asca) capable of high spectral resolution also observe X-rays.

Satellites have looked for evidence of cosmic gamma rays since the early 1960s, but only a few sources have been discovered. The spectacular but still unexplained gamma-ray bursters, sources that appear for seconds or minutes, were detected by military satellites toward the end of the 1960s, but the first reports were published only in 1973. The U.S. SAS-2 satellite (Explorer 48), launched in 1972, and the ESA probe COS-B, launched in 1975, provided maps of the gamma ray emission of the Milky Way and located about 20 individual sources, mostly quasars and pulsars. With the launch of the Russian GRANAT in 1990 and the deployment of the Compton Gamma Ray Observatory (CGRO) from the space shuttle Atlantis in 1991, a new age for gamma ray astronomy has begun.

Two infrared satellites remain to be mentioned, since they provided significant impetus to astronomy and cosmology. IRAS was a joint project by Great Britain, the Netherlands, and NASA. It consisted of a 0.6-m telescope cooled with liquid helium, which in 1983 scanned the sky at four infrared wavelengths: 12, 25, 60, and 100 microns (1 micron = 1/1,000mm). During this mission, almost 250,000 infrared sources were discovered and measured. (Another infrared telescope, the Infrared Space Observatory, was launched by ESA on November 19, 1995. Its spectrographs, photometer, and camera will study in detail the sources discovered by IRAS.)

Launched on November 18, 1989, the Cosmic Background Explorer (COBE) is a satellite designed to study the spectral distribution and spatial fluctuations of the cosmic microwave background due to the Big Bang. The microwave background radiation is the radiation left over from the Big Bang, the first moments of the universe's existence. During its mission COBE became famous for detecting the first details of the formation of the universe, seeing as far back in time as we have yet been able to do.

The potential of space flight for optical astronomy was also exploited before the launch of the Hubble.

But optical telescopes in the past were used almost exclusively to observe objects of the solar system. In general, these were small telescopes aboard probes on missions to planets. The most successful ones include the Mariner probes to Mars, Venus, and Mercury (from 1962), the Viking probes to Mars, and Voyager 1 and 2 to the outer planets Jupiter, Saturn, Uranus, and Neptune (launch in 1977, fly-by of Neptune in 1989); the landing of two Soviet probes on Venus (1970–1982); and the radar mapping of the surface of Venus by the U.S. Magellan probe (1990). In addition, the close encounter of Halley's Comet by the ESA probe Giotto was a spectacular event (1986).

While astronomers at the beginning of the century were confined to the window of visible light, today they are enjoying a panoramic view of the cosmos through numerous windows, which scientists of previous generations did not dare to dream about: the universe can be observed from the longest radio waves to the energetic gamma rays (the exception is the area of extreme ultraviolet, which is absorbed by the gas distributed over space).

The development of the Hubble has opened yet another window. In the area of the visible light we have stepped out onto the veranda of our earthly house – a seemingly small step for science, but it was connected to decades of planning, unimaginable resources, and high risk. And indeed it appeared for a long time as if the skeptics had been right. But slowly – we have to tell the story in order.

The Hubble's Thorny Path into Space

Why an Optical Telescope in Space?

We have seen that telescopes in space are needed to collect and analyze radiation that cannot reach the Earth's surface, and that is an important part of Hubble's mission. But what purpose is served by a space telescope working in the optical domain? Why is it better than a telescope on a high mountain?

This question warrants a detailed answer. A small part of the answer is that some of the light from a celestial object is absorbed in the Earth's atmosphere. This disadvantage could be overcome by building a ground-based telescope that is a bit larger and thus collects more light. However, a telescope does not collect light only from the object itself, but also from the background – and for large telescopes that background signal increases by the same degree as that from the star. What is the background? For the general observer of the sky, some of this background comes from light pollution, the scattered light of artificial sources – street lights, lighted billboards – that interferes with the observation of celestial objects. Astronomers are therefore drawn to remote areas, far from big cities, where they are safe from this scattered light. But even in these areas another troublemaker exists from time to time: moonlight, which is scattered by dust particles and droplets in the atmosphere, causing a brightening of the sky. So astronomers prefer the clear, thin air of the mountains, and they observe the faintest objects only when the moon is not visible.

But even then the sky is not absolutely dark. In the uppermost layers of the atmosphere, the ionosphere, electrically charged atoms combine with electrons and thereby emit certain spectral lines. This airglow may be stronger or weaker, depending on solar activity. At its very darkest, the night sky is as bright as if it contained a star of twenty-fifth magnitude in each and every square arcsecond.

With a telescope in orbit, this airglow can be avoided. Only a much fainter glow, emanating from the gas and dust in the solar system, then constitutes the "natural" sky background.

Another significant factor is the turbulence of the air, which astronomers call "seeing." You can see this effect by looking along a hot blacktop road in the summer – the air shimmers, blurring your view. When a telescope magnifies the view hundreds of times, any shimmering is also magnified hundreds of times. This smears out fine detail. The theoretically nearly point-like images of stars become blurry little disks, which usually have a diameter of between 0.5 and 2 arcseconds. A telescope outside the Earth's atmosphere remains unaffected by this turbulence; there, the diameter of stellar images are governed only by the theoretical resolution of the telescope's optics, which can easily be 20 times sharper. Even in space, infinitely sharp detail cannot be seen. Due to the wave-like nature of light, diffraction occurs at the entrance aperture of the telescope, and a point source such as a star appears at the telescope focus as a very small spot of light surrounded by a series

How to Measure Quantities of Light

Let us look at a simple example: From a star, 100 light particles arrive at our detector on a given telescope during each minute of exposure time. Statistics tells us that we will not register this value of 100 all the time: once it could be 90, once 100, and once 110 particles which are registered by our detector. If we made only one measurement of 1 minute, the value of 100 would have a "statistical error" of 10, in other words, the measurement is only accurate to $10/100 = 0.1 = 10\%$.

Let us assume that our detector is one of the CCD chips used in astronomy, with a million pixels. We have assumed that the light of the star would always fall exactly on one of these pixels. Let us now "switch on" image jitter, so that the image moves around and illuminates not one but four neighboring points – with a quarter of the signal each. We then register a value of about 25 in each of the pixels, but with a statistical error of 5 each. Adding up the values from the four channels, we get 100, but only accurate to 20%.

So far, we have assumed that the background is absolutely dark, but this is in general not the case. Let us look at the light from the identical star, but on top of a background with a count rate of 100. If the signal comes from only one pixel, we measure a total count rate from the star plus background of 200 ± 14, subtract the background (assuming it can be measured accurately), and get a result of 100 ± 14, that is an error of 14%.

If we take image jitter into account again, things get interesting. The background remains at 100, but the signal from the star is reduced to 25 in each of the four pixels, so we measure about 125 ± 11, a total signal from the star of $25 + 25 + 25 + 25 = 100$, and an error of $11 + 11 + 11 + 11 = 44$, or 44%!

So we realize that for faint, point-like objects the image jitter and the sky background are both important for the accuracy of the measurement. If we now assume a signal of 40 instead of 100 for the star, and repeat the above calculations (1) with an image concentrated on one pixel and a background of 1, as we could find in space, and (2) with a smeared or jittery image in four pixels, and a background of 100, as we would find from the ground, we find:

(1) Total signal 41, error 6.4, signal 40 ± 6.4, error of 16%, and

(2) Total signal 4×110, error 10.5 each, signal 40 ± 42, the total error is larger than 100%, larger than the signal itself!

While the first signal (from space) can be measured quite well, its existence itself is questionable in the second (ground-based) case. One would have to integrate almost 100 times as long, to achieve a comparable accuracy in the second case.

of circles of rapidly decreasing intensity. The smaller the telescope, the broader this spot and the more extended the system of rings becomes, and therefore the ability of the telescope to see the tiniest details – its "resolution" – decreases.

Imagine two stars that are very close together in the sky. If the spot image of the second one is distinct from the spot image of the first one, the two sources can be distinguished. A telescope with high resolution, produced by a large mirror diameter working outside the turbulence of the atmosphere, must be the dream for astronomers.

A telescope's theoretical resolution is roughly the product of wavelength and focal length, divided by the diameter of the main mirror or lens. The Mount Palomar 5-m telescope should, in theory, produce stellar images 1–2 microns in diameter, corresponding to a few hundredths of an arcsecond. In reality, the images are almost 100 times worse! Large telescopes on Earth can get nowhere close to the theoretical limit of resolution. Only a telescope in space with comparable dimensions and optics of sufficient quality could deliver stellar images of a few hundredths of an arcsecond, and it should do so not only for visible light, but also for the ultraviolet and near-infrared part of the spectrum. The resolution of Hubble at ultraviolet wavelengths is about three times better than at infrared wavelengths, which are about three times longer.

But astronomers are not only interested in pretty pictures with high resolution. They also want to measure exactly the amount of light the stars emit. When we have to deal with a practically point-like source, the accuracy of these measurements can be dramatically increased with a space telescope. The insert on page 25 explains how unreliable photometry can be with an earthbound telescope.

The deployment of a telescope in space, therefore, has significant advantages for astronomers: terrestrial interference does not play a role; images have a much higher resolution or sharpness; and a space telescope is ideally suited for measuring intensities of faint objects, since faint stars clearly stand out against the much darker background. An impressive demonstration of these advantages is given by the comparison of ground-based and Hubble images on pages 58–59.

From the Idea to the Countdown

Given the great advantages of a telescope in space, it is not surprising that such a project was first conceived long ago. The earliest roots of the Hubble project can be found with the German pioneer of space flight Hermann Oberth during the 1920s, and then much more specifically in 1946, when the American astronomer Lyman Spitzer proposed a large telescope circling the earth – this at a time when it had been known for only eight years that stars derive their energy from nuclear fusion, and for barely twenty years that the Milky Way was not the only galaxy and that the universe was expanding! "The most

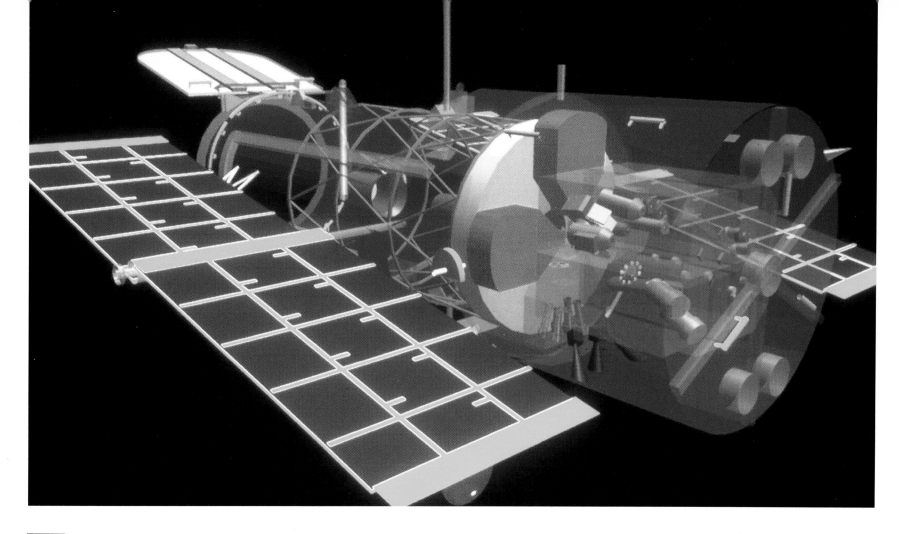

Schematic image of the Hubble Space Telescope, seen from the side and rear: light from celestial objects enters the tube from the left, is reflected by the main mirror (light gray) to the secondary mirror, and comes to a focus after passing the central hole in the main mirror. The Fine Guidance Sensors (red), important for positioning the telescope, and up to five scientific instruments share the light from the telescope. The huge solar panels are to the left and right of the main body of the telescope; they provide electricity for the satellite (Source: ESA).

important contribution of such a radically new and more powerful instrument," Spitzer wrote, "would not be to augment our current ideas about the universe we live in, but to discover new questions which nobody can yet imagine." Fittingly, Spitzer, still working as Professor Emeritus at Princeton Observatory, was one of the first to use the Hubble telescope. With the advent of space flight, the time had come for more concrete discussions. During the 1960s the idea of a large space telescope (LST) was discussed on and off inside NASA and among American astronomers. Lyman Spitzer's voice, 20 years after his visionary words, still carried a lot of weight, and it was he who in 1969 called for a telescope of 3-m diameter in earth orbit. The idea found fertile ground. In 1972 NASA entrusted the Marshall Space Flight Center

with the leadership of the project. Robert O'Dell was appointed project scientist for LST, and thus became the father of the project (together with Spitzer as grandfather and Oberth as great-grandfather). The Marshall Space Flight Center was glad about the new mission, because after the abrupt end of the Apollo program it was looking for a spectacular project similar to the landing on the moon. However, funds to build and launch the telescope were not secured for a long time.

Also in 1972, NASA started to develop the first reusable space transport system, the Space Shuttle. From the beginning, the fate of the Space Telescope was tightly coupled with that of the Space Shuttle. The shuttle would not only deploy the telescope in orbit, but also make regular visits. The telescope

was to be docked with the shuttle, and astronauts were supposed to enter the rear part of the satellite to exchange instruments. The promise of regularly scheduled visits by very inexpensive shuttle flights was supposed to reduce the cost of the Space Telescope as well. Regularly scheduled maintenance was to permit the Space Telescope to "achieve the immortality of the great telescopes on earth," as O'Dell promised in 1972. The astronomers believed that NASA would fund its development beginning in 1976, and launch was expected by 1980.

But between the dreams of astronomers and space stood the U.S. Congress. After the oil crisis of 1974, the space telescope project almost ended before it began. Congress was interested only in the cost of the project, and $400–$500 million over six years was too much. Deliberations took less than five minutes – and then all funds were cut from the budget! Only now did astronomers realize that not everyone shared their euphoria for "great science." They had neglected to advertise their project, and not even all astronomers supported it.

Again it was Lyman Spitzer who rallied the troops. A miracle took place: Spitzer succeeded in generating a resolution, which was also supported by previously LST-critical astronomers, according to which "the Large Space Telescope has the leading priority among the future instruments of space technology." Congress was lobbied with one voice – and opinions changed for the better. Originally, $6.2 million were

supposed to be approved for further studies in 1975, but President Ford, who was emphasizing austerity, cut this amount in half and demanded that NASA seek the "substantial participation of other nations." The astronomers had won, but the telescope, which was funded by NASA beginning in 1977, had become smaller (the mirror diameter decreased to 2.4 m) and it would be launched later. The main contractor for the satellite was Lockheed in California, which had earned its reputation by building high-resolution

Top left: An ESA manager explains a full-size model of the Faint-Object Camera (Source: D. Fischer). Top right and bottom: The solar panels are large even when fully retracted. Unfurled (top), they are so flexible that they have to be stored on water during tests (Sources: D. Fischer and ESA).

Hubble shortly before completion. Note the retracted solar panels (gold) and the high-gain antenna (black) – and how the telescope dwarfs the technicians below (Source: NASA).

reconnaissance satellites. The Hubble telescope was supposed to profit from this technology. The heart of the satellite, the optics, were to be built by Perkin-Elmer in Connecticut, a company that had succeeded against two competitors with a very low bid and the argument that their management experience was particularly good. This assessment turned out to be a grave mistake.

Was the Hubble mission to be solely a NASA project? Cooperation with the Soviet Union was unthinkable at that time, and the Japanese space program had not progressed far enough. But the Europeans were interested. Since 1973 there had been contacts between NASA and the European Space Agency (ESA) about a potential participation, and this was favored by the United States for financial reasons. This participation finally consisted of the following items. First, ESA provided the Faint-Object Camera (FOC). The technology for such a camera, which registers every particle of light as a function of position and time, was most advanced in Great Britain – it was Europe's ticket to the Hubble. At the University College London, such an instrument had been developed for ground-based telescopes, and an improved version was planned for the Space Telescope. Second, ESA provided positions for European scientists at the Space Telescope Science Institute in Baltimore, and established the Space Telescope European Coordinating Facility in Garching, near Munich, Germany. Finally, ESA provided the critical solar arrays that generate electricity for the Space Telescope. Subsequently, a memorandum of understanding was established between NASA and ESA in 1977, providing ESA approximately 15 percent of the available observing time with Hubble in exchange for its support.

At this time things happened that have happened with other large-scale projects. During the construction phase all financial plans had to be thrown out the window, as the project cost grew to well over a billion dollars, four times as much as originally planned. The first concrete launch date envisaged by NASA was in October 1983. However, the complexity of establishing a large astronomical observatory in low orbit had been severely underestimated, and technical problems forced several postponements.

At the beginning of 1986, launch seemed but a few months away. Then the unthinkable happened. On January 28, 1986, shortly after lift-off, the space shuttle *Challenger* exploded, killing the entire crew. The shuttle program was suspended for two and a half years. Like nearly all other space projects, Hubble had to be postponed once again. The complete telescope was stored in a very large, very clean room at Lockheed for four years. The solar panels were removed and brought back to Europe. By the end of 1988, shuttles were flying again, and the Space Telescope, considered one of the most important civilian projects, was scheduled for launch in April 1990.

According to plan, the countdown for the space shuttle *Discovery* began on April 7. Previously, the Space Telescope had been brought to Cape Canaveral on a huge transport plane. Once installed in the cargo bay of *Discovery*, it became the most expensive civilian payload ever launched into space. Steadily, the April 10 launch was approaching. Thousands of engineers and technicians had devoted years of their lives to this project. For them, and for the astronomers, a dream years in the making was ready to come true. After all, wasn't their problem child, the Space Telescope, securely attached to the payload bay and fully operational? One or the other of the scientists and technicians, having followed the development for years, might have had nagging doubts – but this was not the time for them. Until nine minutes before launch everything happened according to plan. The last "hold" during the countdown, the last planned check before lift-off, did not reveal any problems either. But still, five minutes before launch, as the auxiliary power units, which operate the orbiter's hydraulics, were switched on, there were indications that they were not running smoothly.

Launch director Bob Sieck let the countdown continue, but at the very last minute, when Hubble's historical journey was supposed to begin, he had to cancel the launch. The approximately 200 relatives and descendants of the astrophysicist Edwin Hubble, who wanted to witness the launch at the Kennedy Space Center, had to leave in disappointment.

Exactly two weeks later the defective auxiliary power units had been replaced in record time. And

Lift-Off! On April 24, 1990, at 8:33 A.M. local time, the space shuttle *Discovery* with five astronauts and the valuable payload on board rises above the launch pad at Cape Canaveral, Florida (Source: NASA).

again the countdown got under way. On April 24, 1990, at 8:30 A.M. local time, it had reached T − 60 seconds, and this time all three auxiliary power units operated flawlessly. At T − 31 seconds, control of the countdown is supposed to be transferred from the control center to the shuttle itself – but *Discovery's* on-board computer believed a valve configuration to be wrong, and stopped the countdown. Immense tension set in, as the hydrazine fuel of the auxiliary power units will only last for a few minutes. But in the shortest possible time the valve status message was identified as a software error, and the countdown continued. At 8 hours, 33 minutes, and 59 seconds A.M. EDT, the space shuttle *Discovery* lifted off the launch pad, with the Hubble Space Telescope, "our window to the universe," on board, as the NASA launch commentator called it in an undoubtedly well-rehearsed phrase. *Discovery* disappeared briefly in a thin cloud, and then became visible again – and attained an altitude of over 600 km only 8 minutes later. The launch was successful. For 15 years Hubble is supposed to be this window to the universe, and should help astronomers solve numerous questions. Rarely have there been such great expectations at the launch of a satellite – and rarely have they been, at least in the first years, so bitterly disappointed.

In Orbit

Only a few times had humanity been as far from Earth's surface as the crew of *Discovery*, which deployed the Hubble at the highest altitude it could reach. The last time anyone had been so high was during the final Apollo moon mission, 18 years earlier. Every kilometer gained in altitude for the Space Telescope deployment will prolong its life in orbit, because even at this altitude of over 600 km there is enough residual atmosphere that this huge satellite is slowed down and consequently gradually descends. At least once during its expected 15-year lifetime, Hubble will have to be boosted into a higher orbit with the shuttle (the satellite has no rocket engines of its own) – a complicated and risky maneuver. The higher Hubble starts its life, the better. The nearly circular orbit of *Discovery*, attained a few hours after launch, is at an altitude of between 613 and 615 km. Preparations for Hubble's deployment can begin.

Five hours after launch, the satellite transmits its first signals, but *Discovery* is still providing the power. One day later, on April 25, the "umbilical" is disconnected .Only eight hours remain to deploy the solar panels, so that the telescope's batteries do not drain too much. Fully unfurled, they are 12 m long and extremely thin. Without support they would collapse under earth's normal gravity, so they and their 48,760 individual cells can only be used under the weightless conditions in orbit. For launch they fit into the payload bay, rolled up in canisters and stored on the sides of the telescope. All solar cells combined generate about 4.4 kW of electrical power for the satellite and all its instruments.

Deployed! This fascinating picture was taken seconds after the release from the robot arm of *Discovery*. Hubble's still closed front cover reflects the Earth (Source: NASA, Smithsonian Institution, and IMAX Corp.).

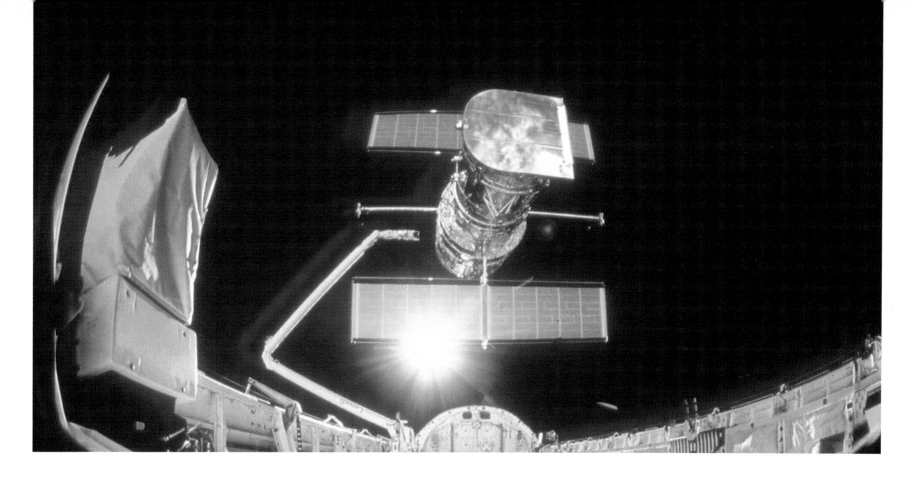

In the morning of the second day, the Space Telescope is lifted out of the payload bay, with the help of the Canadian remote arm with astronaut and former astronomer Steve Hawley at the controls. The communications antennae are deployed, and the solar panels unfurled – but one of the panels is stuck. This problem is considered serious enough that two astronauts begin to prepare for a space walk in order to help the panel along. The astronauts have trained extensively for such emergency intervention, but in the end the space walk is not necessary: the panel comes unstuck on its own. With solar panels fully unfurled and battery charge increasing again, and with high-gain antennae deployed, the entire telescope is released at 3:38 P.M. EDT. Hubble is now a true satellite. *Discovery* moves away very slowly, to avoid last-minute contamination of the satellite by its rocket thrusters, which might ruin Hubble's painstakingly maintained clean-

liness and degrade the ultraviolet reflectivity of the optics.

Aside from occasional glances at the telescope through their binoculars, the astronauts did not have anything to do with the Hubble any longer, but they remained at the ready for two more days; in case of severe problems they could have captured the satellite and brought it back to the ground. Nothing could have been more embarrassing to NASA than to leave a nonfunctioning telescope in orbit, and therefore the astronauts had trained for all sorts of emergencies – for instance, if the huge front cover of the telescope failed to open. Indeed opening the cover took several hours because of communications difficulties. This procedure put the satellite for the first time into "safe mode." Whenever it senses danger, Hubble's on-board computer makes itself independent of most commands from the ground and puts the satellite into an attitude where the solar panels receive enough

The Wide-Field Planetary Camera

The Wide-Field Planetary Camera, or WFPC (pronounced Wiffpick), and its successor, WFPC-2, are the only instruments of the Hubble – aside from the Fine Guidance Sensors – that access the optical path from the side. In size and shape they are comparable to a baby grand piano. The old WFPC actually consisted of eight cameras, with four of them active together, depending on the position of a small glass pyramid deflecting the light to them from the Hubble optics. Four CCD chips in the four cameras, arranged in a 2 by 2 square, were illuminated together. When Hubble and the first WFPC were conceived, these Charge-Coupled Devices were at the cutting edge of detector technology; today they can be found in many camcorders. Hubble's CCD chips had 800×800 picture elements, or pixels. In those, electric charge was collected proportional to the light intensity arriving at the respective points of the chip, and then read out and transmitted to Earth. In principle, the WFPC chips were supposed to be sensitive from 115 to 1000 nm, but contamination on the chips, which had formed in spite of all efforts to keep them clean, somewhat reduced the sensitivity at wavelengths shorter than 300 nm – the entire ultraviolet range.

The area covered by the four chips of the Wide-Field Camera measured 2.7×2.7 arcminutes – for a mosaic of the moon, although not possible because of its brightness, almost 100 images would have been needed – with a resolution of 1/10 arcsecond per pixel. The Planetary Camera covered a field of 1.1 arcminutes on a side, but with more than twice the resolution (0.04 arcseconds). In addition to distant galaxies, the largest planet, Jupiter, would just fit into its field of view, hence the name. A total of 48 color filters and polarizers were available to isolate part of the spectrum and even single spectral lines. In principle, the camera electronics allowed exposure times between a fraction of a second and several hours, but the practical limit was about one hour. Stars down to 28th magnitude were expected to be observable with the WFPC – objects a hundred times fainter than available to the ground-based telescopes and instruments of the 1970s and early 1980s. Around the time of launch, however, improved optics and better electronics had opened up similar limits for ground-based equipment. The gulf between ground-based and space astronomy had begun to narrow, just as Hubble became operational. However, Hubble's combination of acuity and sensitivity (down to ultraviolet wavelengths) cannot be achieved from the ground.

light, and the telescope points away from the sun. In this configuration the telescope is protected. Hubble is not brought out of safe mode until the problem that put it there is understood, which can take hours or even days.

The technical difficulties did not stop. A particularly serious issue was the disturbances that occurred every time Hubble crossed the day-night boundary in its orbit. It quickly became clear that this was due to warping and deformation of the huge solar panels. This sometimes caused the satellite to lose its lock on its guide stars, which were expected to provide pointing to an accuracy of a tiny fraction of an arcsecond. The finely tuned gyro systems, which were to be used to move the Hubble in all three axes, were too weak to counteract the shaking. Thus one of the great mishaps in Hubble's construction began to emerge by mid-May, a fundamental and not completely correctable design flaw, which threatened to hinder the scientific work considerably. At this time no one suspected the optical problem that later would be major news.

What was going on with the actual work of the Hubble? One week after launch, things were four days behind the ambitious plan, which called for a so-called orbital verification phase of three months, followed by a five-month scientific verification phase for the instruments to be calibrated and scientific procedures tested. At the end of 1990, the first guest observers (regular astronomers from the U.S., Europe, and elsewhere) were supposed to begin receiving observing time. Before then, use of the telescope was reserved for the scientists of the teams providing the scientific instruments. About ten times more observing time had been requested than was available, and it was expected that the demand would grow even further.

But none of the scientific instruments was operational as yet. First, the focus of the entire telescope had to be determined, and the optical elements of the Hubble had to be adjusted accordingly. The deviation from perfect focus of images of pointlike stars was being measured by the so-called fine guidance sensors, the instruments used to perform the accurate orientation of the Hubble. But to do so, sufficiently bright stars had to be found, and this occurred only on May 4, after several additional days of head scratching, due to an error in the calculations. After several movements of the secondary mirror, the position of the focus appeared to be good enough to venture the first image. NASA had organized a huge media event around this "first light," as astronomers call a new telescope's first exposure. The press had demanded that the taxpayers, who had invested so much money, be presented results as soon as possible after the television show of launch and deployment. NASA had not clearly explained how long a new major telescope and instruments like this usually take to start up, and had committed itself to providing images much sooner than originally planned. The experts knew that at this early in the mission, the images would hardly be sharper than what was obtainable from a good ground-based telescope.

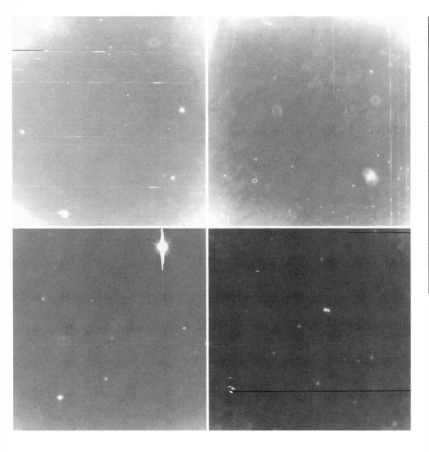

GROUND BASED IMAGE
LAS CAMPANAS OBSERVATORY
CARNEGIE INST. OF WASHINGTON

HUBBLE SPACE TELESCOPE
WIDE FIELD/PLANETARY CAMERA

NASA

Ground-based Image
Nordic Optical Telescope

Faint Object Camera
Hubble Space Telescope

NASA ·e·esa

Historical images: "first light" for the Wide-Field Planetary Camera on May 20, and for the Faint-Object Camera on June 17, 1990. The larger image on the left shows the original data of the 30-second exposure with all defects and contaminations: the images of stars are relatively concentrated, which led NASA to publish a cutout, comparing it to a ground-based image (top right). The telescope's optical flaw had not been discovered at that time. A similar comparison was done for the first-light image of the Faint-Object Camera (bottom right) (Sources: ST ScI, NASA, and ESA).

On Sunday morning, on May 20, 1990, the "first light" image with the Wide-Field Planetary Camera (WFPC) was supposed to be taken, and NASA was hoping for a publicity splash at the brightly illuminated Goddard Space Flight Center, from which the actual commands were sent to the spacecraft.

At 6 A.M. local time, the commands for taking two images had been uplinked to the satellite. A star cluster in the Southern hemisphere, NGC 3532, was the target object. More spectacular targets, like the Eta Carinae complex, could not be used because individual scientists had claimed "proprietary rights" on observations of many objects, and there was little agreement on the importance of the timely presentation of some objects of interest to the public.

The first image was exposed for only one second and the next for 30 seconds, to be sure to get a good exposure. At first there was only the confirmation that the shutter of the camera had indeed opened and that the data had been written onto the on-board tape recorder. Transmission to Earth would take some time. Since Hubble's orbit is low enough that it can directly send signals to only a limited area of the ground at any one time, communication requires a relay satellite. Hubble has to share this satellite with several civilian and military U.S. satellites, and wait until it is in the correct position and a free channel is available. In the meantime, data are stored by the on-board tape recorder. Finally everything was ready, and the transmission of images began: from Hubble to the relay satellite, from there to a ground station and up again to another communication satellite in geostationary orbit, and then to the computer systems at Goddard and Baltimore. Even at the speed of light this takes a few seconds.

The first image, with an exposure time of one second, did not show much but one relatively bright star, apparently about half an arcsecond in image size, surprisingly good considering the early time after launch. The second image, with longer exposure time, showed several more stars – the camera was working, and apparently the optics as well. The adventure had begun. But already these first pictures revealed something strange about the stellar images. Instead of a symmetrical distribution of light, they consisted of a central spot plus a blurred disk of over an arcsecond

in diameter, which for the brightest stars revealed structure reminiscent of tentacles or spider legs. Such a composite shape of the stellar images had not been observed at any ground-based telescope working correctly, and it had not been expected from the Hubble either. During the following weeks, as more images showed the same mysterious shape, concerns began to grow, and suspicions about a fundamental problem arose. Were the telescope's mirrors perhaps significantly less smooth than indicated by final tests on the ground? Had they been damaged on their way into orbit – or was there a major construction flaw?

The public had been informed about many of the troubles plaguing the Hubble during its first weeks, but the strange shape of the stellar images was not revealed. During May and June several images were published, but never in a quality that showed this bizarre structure. Only the sharp central parts

The Faint-Object Camera

The Faint-Object Camera (FOC) applies a different principle than the WFPC. It registers every light particle individually. The FOC consists of two identical photon-counting detectors, each made up of a three-stage image intensifier (with an intensification factor of 100,000) and a video camera with image processing electronics, determining the location of the light particle on the detector. This technique gives the FOC enor-

mous sensitivity, but it cannot look at brighter sources: The detectors cannot cope with more than one photon per pixel per second. Furthermore, the FOC is much more sensitive than the WFPC in the ultraviolet range, and it also has much better angular resolution. In one of the two channels, the pixel size is 0.04 arcseconds, with a total field of view of 22 arcseconds, while in the other one, pixel size is 0.02 arcseconds, with a total field of view of 11 arcseconds.

of the images were seen, not the fainter strange detail. Was the world, including the astronomers, misled on purpose, as some critics would later accuse? The main reason for delay was simply that it took astronomers and engineers quite a while to convince themselves that such an unexpected problem existed. And caution was justified; had NASA publicized the flawed images right away and speculated loudly

Hubble's Optical System

Hubble's optical system is of the Ritchey-Chrétien type, which had been used for ground-based telescopes for some time. The advantage of this configuration is a relatively large field of view with images virtually free of residual optical distortions. The drawback is that both the primary and secondary mirrors are in the form of a hyperboloid, a shape that is difficult to achieve during the grinding and polishing process. The main or primary mirror, a lightweight honeycomb structure of titanium silicate glass, has a diameter of 2.4 m and a focal length of 57.6 m. Its total mass is about one metric ton. The secondary mirror has a diameter of 0.34 m. It reflects the light through a central hole in the main mirror in a focal plane, where the scientific instruments and sensors for positioning the telescope are located. Since the telescope is not inhibited by Earth's atmosphere, special requirements on the accuracy of the mirror surfaces had to be imposed to approach the theoretically possible resolution of the telescope. An accuracy of 1/20th the wavelength of light over the entire system was attempted, requiring an accuracy of 1/50 for the main mirror, and 1/110 for the secondary mirror. The grinding and polishing of the mirror surfaces was done by computer-controlled equipment. After each step, the mirror surfaces were analyzed using laser interferometers. At the end, an accuracy of 1/90 was achieved, surpassing the goal. Finally, aluminum was evaporated onto the mirror surfaces to give them the needed reflectivity, and a protective coating of magnesium fluorite was applied. The extreme precision required for the mirror surfaces was new, but achieved and even surpassed in every aspect. Note that the much-publicized flaw of the mirror has nothing to do with the smoothness and quality of the surface – it was just manufactured to the wrong shape, but very precisely so.

The mechanical assembly of the satellite is centered around a large main ring of titanium. On the front side, the mounting of the telescope, providing support for the mirrors, and made from carbon-fiber-enforced synthetic materials, is attached. The back contains the bays for the scientific instruments. Electronics and equipment for data storage and communications are attached to the central part of the satellite.

about the reasons for them, the various companies involved would have had a justified complaint. First it had to be established with reasonable certainty where in the complex optical system the imaging flaw occurred.

Because Hubble's positioning still posed a few problems, the satellite had not been moved out of the Carina region in the southern sky – the controllers at the Goddard Space Flight Center were afraid to lose the orientation completely by performing a larger telescope motion. It was still impossible to routinely maintain lock on guide stars with the fine guidance sensors, so that exposure times of more than about a minute led to blurred images. The entire satellite still vibrated for almost ten minutes after each crossing of the day-night boundary. It was clear that the construction of the extremely delicate solar panels contained a severe fault.

By June 17, Hubble's second camera, the Faint Object Camera, was ready for its "first light" as well. The open star cluster NGC 188, near the celestial North Pole, had been selected as its target. Seven exposures were successfully taken during the morning of June 17, and they showed the expected stars – but again in strange-looking images, with a bright center and an extended halo of light containing the unfortunate "hairy tentacles"! Now it was clear that the flaws could not have been introduced by the WFPC. Hubble's main optics had to be defective. The telescope itself, the optical foundation for all the instruments, was flawed.

The Scandal of the Defective Mirror

The scandal erupted. Obviously, one of Hubble's mirrors, probably the primary mirror, deviated significantly from the correct shape. According to all indications it was too flat by 2 microns, only a few percent of the thickness of a hair, but devastating by optical standards. The diagnosis was spherical aberration. This optical fault is easily explained: An ideal telescope unites parallel rays, as they come from a point out in the depth of space, in a single point in the focal plane. When aberration occurs, such a well-defined focal plane does not exist. Some light is focused there, but the majority ends up either in front of or behind the plane, the result being the extended halo of light surrounding every stellar image from the Hubble. Now NASA abandoned its reserve. It was inevitable that the debacle had to be communicated to the world.

On June 26, NASA called a special press conference for the next afternoon; the invitation mentioned that the secondary mirror had been moved into all possible positions, but "the expected image quality had not been achieved," and "computer modeling of the images points to spherical aberration as the cause." Of course there were questions as to the origin of the optical flaw at this press conference, and even without a detailed examination it was correctly assumed that the complexity of the mirrors was the reason the flaw was not detected. In their defense, the project managers pointed out that the combined test-

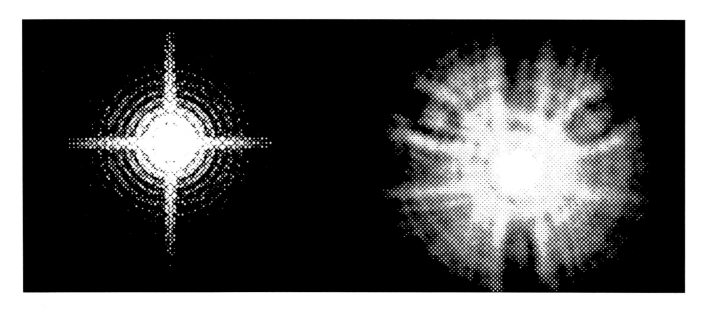

Wishful thinking and reality in Hubble's pictures of 1990: Stars were supposed to look like the computer simulation on the left, but instead they resembled the one on the right – "squashed spiders," as they were soon called. Experts on optics could immediately conclude from such images that there was a severe problem (Source: Allen Report and ST ScI).

ing of both mirrors using artificial stars would have required several hundreds of millions of additional dollars. Furthermore, there was the danger of contaminating the optical surfaces – not to mention the technical difficulties of testing such a huge instrument, expected to work in the weightlessness of space, on the ground. So the project relied on separate tests of the main and the secondary mirror – a severe mistake, as the managers now had to admit.

This disastrous press conference, coming on the heels of gloomy assessments by various groups of scientists, would color the public image of the Hubble telescope, and that of NASA, for years. The agency has often been accused of having discounted the value of the billion-dollar instrument more than the spherical aberration alone would justify, but that is incorrect. What NASA knew at that time, that many of the scientific programs could be salvaged, was not accepted by the media because proof, in the form of spectacular images, did not yet exist.

The main effect of Hubble's vision problems was that the Space Telescope was not able to focus the majority of the incoming light into a single small point. Specifically, 70 percent of the energy of a stellar image was supposed to be concentrated within a radius of 0.1 arcsecond. But now only 10 to 20 percent could be so concentrated. That could have been tolerated, if the remaining 80 or 90 percent were not distributed over the surrounding arcsecond or more. At first glance it gave the impression that Hubble could see no better than any telescope on the ground. Had there been a Hubble image of, for example, a dense star cluster at that ominous press conference, it would have been obvious that the Hubble could still approach the planned image *resolution*, being able to separate stars only 0.1 arcsecond apart or less. The aberration mainly caused a reduction in *sensitivity*, because only 15 percent of the light was focused in the sharp inner core of an image, and a loss of contrast. Hubble was neither blind nor near-sighted, as it was frequently described, but it saw the universe as if through fogged glasses. Already at the time of the press conference there were efforts under way at the Space Telescope Science Institute to remove the fog via computer, but these techniques were not fully tested at that time due to the lack of suitable data, and many traditional astronomers viewed them with great suspicion.

Images like this one, where
the telescope was deliberately
defocused, were helpful in
discovering the spherical
aberration. The shape of the
"squashed spider" changed in
a characteristic way (Source:
ST ScI).

Was the Hubble still usable for science, or would the entire project have to be written off? Many people maintained that its usefulness would still far outweigh the damage. By comparison, the telescope's capabilities were affected only slightly for spectroscopy (the dispersion of light into its component colors, performed by two of its five science instruments) and photometry (the measurement of stellar brightness), and not at all for reaching into the ultraviolet range of the spectrum – one had only to expose a little longer. Opinion among scientists was clear: Several of the imaging programs would have to be postponed, and work should focus mainly on spectroscopy and ultraviolet observations. In addition, it was already foreseeable that the spherical aberration could be characterized so accurately that new scientific instruments could correct for it. They would be installed during one of the Space Shuttle's planned visits. These regular visits, which would also be used to exchange other elements of the satellite, now became the great hope of the project. The mission would be restructured so that in two or three years a new Wide-Field Planetary Camera with built-in image correction could be installed, and everything currently not feasible could be done then. A faint consolation.

Yet again a seemingly trivial fact had almost brought down a billion-dollar project, and NASA with it. The parallels to the tragic loss of the Space Shuttle *Challenger* four years earlier were all too obvious. Then the brittleness of a cold rubber seal had been the cause of the disaster, now it was a defective mir-ror. However, the underlying causes in both cases had been management mistakes and questionable developments within NASA itself. The Space Telescope project, intended to be a crowning achievement of American space exploration, was now at the center of severe criticism. How did all this happen?

Few people were as familiar with the Hubble project from the very beginning as Riccardo Giacconi, director of the Space Telescope Science Institute from its beginnings in 1981 through 1993. To him, the idea of a low orbit, though reachable by the Space Shuttle, was questionable. In a higher orbit, the telescope could have been operated much more effectively and conveniently. A second point of criticism was the low "local intelligence" of the satellite, requiring the communication of every little operational detail from the ground. All these shortcomings existed because astronomers had never been asked. NASA considered the Space Telescope its own project from the very beginning. The agency wanted to maintain control of all aspects of the program and therefore would not delegate the management of the Space Telescope to an external scientific institution, as the National Academy of Sciences had requested.

Only the "intellectual" part of the project, the planning and execution of the science program, was eventually left to the astronomers, who settled down at Johns Hopkins University, in Baltimore, Maryland – close enough to the NASA Goddard Space Flight Center, in Greenbelt, Maryland, to have access to the data stream from the telescope, but removed from

Riccardo Giacconi, first director of the Space Telescope Science Institute. He is one of the pioneers of X-ray astronomy, and participated in the discovery of the first X-ray source outside our solar system (Source: ST-ECF).

the space agency's direct access. The scientists were allowed no say in the construction of the satellite itself, not even in the development of the software for their own computer systems, which was given to an industrial contractor. "There was no intellectual leadership, and nobody was really responsible," said Giacconi. The software developer had designed the control of the telescope in such a way that 36 (!) people had to continuously devise command sequences . Only the astronomers had the idea to use artificial intelligence, so that the control laws could be formulated at a higher level, with the computers taking care of the detailed command sequences.

It took several years to improve the control software, which had been delivered in 1985. Project management had particularly underestimated the problems related to the extremely accurate positioning of Hubble in space. To keep the satellite stable during an observation it was not enough to rely on the gyroscopes alone. A pair of guide stars had to be acquired and "held on to" with the fine guidance sensors. However, the positions of these stars had to be known before observation could begin, and the best existing star catalogs contained only the several hundred thousand brightest objects. NASA had thought it would be enough just to get a photographic plate of the respective area in the sky from the archive on the day of the observation, and measure stellar positions around the target object to an accuracy of 0.25 arcsecond! To provide enough guide stars, the scientists in Baltimore decided to create the "Guide Star Cata-log," a gigantic catalog of over 15 million stars – one of their first great achievements.

The more the people in Baltimore became acquainted with the intricacies of Hubble operations and of dealing with its data, the larger became their worries before the launch: Could there be flaws in the system, which we do not know about? These questions were posed almost at random – but today we know: Had there been enough time to follow up on the question of whether or not the mirror system could deliver correct images, astonishing details would have been discovered. On the day of the depressing press conference of June 26, 1990, an investigating commission under the leadership of the director of the Jet Propulsion Laboratory, Lew Allen, had already been formed. Within a few months he presented a report containing clear statements and precise conclusions.

"During the years 1981–82 the project was plagued by many problems," the commission wrote. "The estimated cost of the Perkin-Elmer contract had been increased several times, and the schedule had slipped. . . . The complexity of keeping the mirror clean had only recently been recognized. The program was threatened by cancellation, the capabilities of management were in doubt. All these factors appear to have contributed to a situation in which NASA and Perkin-Elmer management were distracted from monitoring the production of the mirrors."

The fact that both mirrors in the Hubble telescope have hyperbolic rather than spherical shape made testing of the individual pieces difficult – it involves the use of so-called null correctors. With this technique, wave fronts are generated by a complicated optical setup in such a way that the mirror to be tested appears as a simple spherical mirror. A proven procedure, but the specialists at Perkin-Elmer considered the usual method of a *"refractive* null corrector" not sufficiently accurate for the quality demands of the Hubble project. Therefore, a novel *"reflective* null corrector" was built. Perkin-Elmer planned to test the reflective null corrector very thoroughly but they planned no independent tests of the mirror itself. The idea of the new null corrector had been one of the reasons Perkin-Elmer won the contract to produce the Hubble mirrors in the first place.

Unlike the classical null corrector, the reflective null corrector can, in principle, measure and re-verify the distances of the individual components at all times, using a so-called "metering rod" of accurately known and very constant length. When the Allen-Commission examined the old test measurements, it made an astonishing discovery: the refractive test had shown spherical aberration of the primary mirror, but this null corrector had been flatly declared defective, since the new reflective test had not shown the aberration! When the commission verified the refractive null corrector, they found it to be perfect. The reflective one had to be in error.

Luckily, the testing apparatus had not been changed since the completion of the mirrors, and the rest of the explanation of the greatest optics scandal in the history of astronomy was simple. A

The core of the problem: Hubble's perfectly misshapen main mirror, here at tests at the manufacturer, Perkin-Elmer in Connecticut (Source: NASA).

small lens was positioned 1.3 mm too far from one of the mirrors, a comparatively huge distance. The metering rod was still available as well, including a small field cap on one end, made from nonreflective material and containing a small aperture. A laser beam was supposed to fall through this hole onto the end of the metering rod, so that the distance to the mirrors could be determined with extreme precision. But a small piece of paint had been chipped off the field cap, and the laser beam had been reflected from there, not from the end of the metering rod 1.3 mm below! The measuring scenario was reenacted, and the identical error occurred promptly. Actually, the reflective null corrector should have been tested independently using different methods, and only then declared optically correct, but other tests obviously had never been done, and a formal declaration of its correctness had never been issued.

The findings of the Allen Commission were clear: "Competent individuals at Perkin-Elmer and at NASA would have had to be alarmed. Everybody should have noticed that the check of the null corrector was crucial, and that it and the mirror should have undergone independent testing." But, amazingly, "there was a remarkable lack of experts for the construction of large telescopes during the production of HST." Furthermore, "the technical advisory group of Perkin-Elmer showed no interest in the production process and, in spite of their knowledge about the error possibilities of the measuring technique, did not voice any concerns or check the test results." There were, after all, known cases of spherical aberration with other telescopes of similar design. The warnings of an advisory panel, pointing out the possibility of a large error and proposing simple tests, had been ignored. Yet a simple test, costing only a few thousand dollars and familiar to any amateur astronomer, would have uncovered the flaw.

Hubble's Difficult Years

During these days and weeks, the future of the entire project was highly uncertain. Aside from the servicing mission, several other possibilities were being discussed. Riccardo Giacconi argued against a repair mission and in favor of building a second satellite. "We take the second mirror, which is still at Kodak, make a new satellite, approach the Russians, charter an Energia rocket, and launch it into a geostationary orbit," he urged.

According to Giacconi's estimates, the cost of a second Hubble telescope would not have been higher than for a repair mission. "The Russians were in favor of it," Giacconi remembers, "but not the Americans. The shuttle *had* to be used, and the concept of in-orbit maintenance had to be demonstrated, to prove that the Space Station was feasible." So that was how it happened, and during the planning and execution of the repair mission at the end of 1993, the various NASA institutions and the independent Space Telescope Science Institute worked together in exemplary fashion for the first time. To correct the optical flaw several possibilities were discussed, including a brutal reshaping of the mirror, the depositing of a new mirror coating, and the mounting of huge corrective lenses or mirrors. Finally, a concept by the name of COSTAR was adopted.

The COSTAR concept was presented on October 26, 1990. It sounded so convincing that NASA changed its strategy and gave out the contract to build the instrument just two months later. Whereas the specter of complete cancellation had been hanging over Hubble only six months before, it now appeared that an almost complete restoration of its capabilities within three years was well within reach.

We should not forget that during this time the telescope delivered scientific data and results. The roller coaster of mood swings had taken a decidedly upward turn during the second half of 1990. After the prelaunch hype and the subsequent disaster, things began to change again. The two spectrographs had

The image that may have saved the entire project: The extremely compact star cluster R136a in the Large Magellanic Cloud, resolved into hundreds of individual stars (Source: NASA).

COSTAR – Hubble's "Glasses"

In order to correct Hubble's optical flaw, two new mirrors per instrument aperture were inserted into the optical train. The first one is a simple spherical mirror, reflecting the light onto the second one. The second one has to correct the spherical aberration of the telescope's main mirror, so that the light arriving at the instrument apertures is free of the optical flaw. The ability to manufacture such aspherical mirrors, albeit very difficult, was of crucial importance. For future instruments, including the WFPC-2, similar correctors can be incorporated into the instrument itself. But how should this be done for the existing instruments? The answer came from the modular design of the instruments: a shell with the dimensions of an axial instrument was used, in which the complicated mechanism for moving the correcting mirrors into the optical train of the Faint Object Camera and the two spectrographs was incorporated.

The complete system was called COSTAR (for Corrective Optics Space Telescope Axial Replace-

ment), and consisted of about 5300 individual parts. On an extendible optical bench ten mirrors of the size of coins were mounted, seven of them on light beryllium arms. To that were added twelve motors to extend the arms and to correct the position of the mirrors, as well as numerous sensors to check the position and temperature (to about one kelvin). One anecdote attributes the idea for the COSTAR principle to an engineer taking a shower in his hotel during a conference, and as it happened, the shower head could be moved up and down as well as tilted – and that is how COSTAR works. The ability to move all components in a controlled way was crucial. The entire mirror system must be removable in case of any malfunction, so that the scientific instrument apertures would not be obscured, and each individual mirror must be freely movable as well so that the correction could be optimized (this is also true for the respective elements of the WFPC-2 and other future instruments). To make room for COSTAR, the least used instrument – the High Speed Photometer – had to be removed from the telescope.

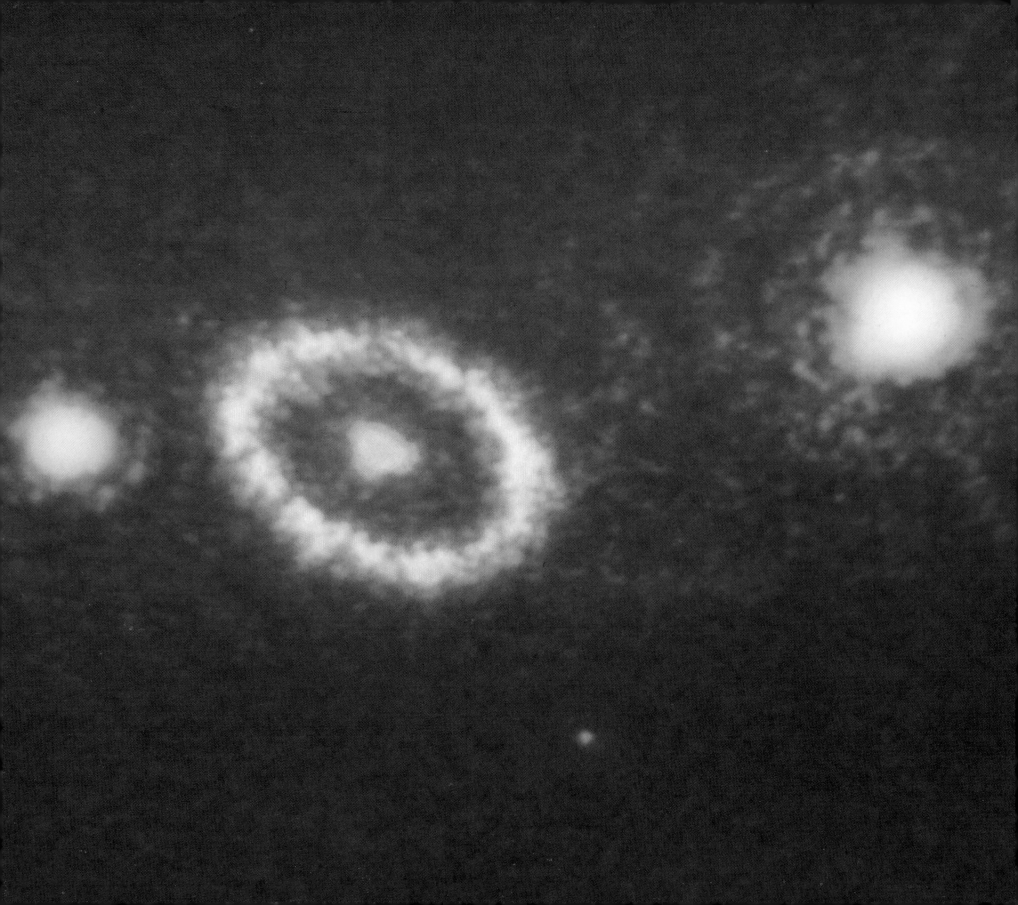

The glowing gas ring around Supernova 1987A, located in the Large Magellanic Cloud as well, and then three and a half years old. Only Hubble's images showed how thin it actually was. Even now its formation remains mysterious (Source: NASA and ESA).

been activated. It is their task to disperse the light from stars or other objects into individual colors, providing information about their physical characteristics and chemical composition. The instruments on the Hubble had two distinct advantages over similar ones on ground-based telescopes: their sensitivity reached far into the ultraviolet, and they could provide spectra of very small areas, for instance within a galaxy or nebula, thanks to Hubble's high resolution and their small apertures. It was during one of the first attempts with a spectrograph that the Hubble delivered its first "real" stellar image.

The target was an object with the prosaic name R136a in a small galaxy neighboring the Milky Way, the Large Magellanic Cloud. Earlier, a single extremely massive star had been suspected there, but during the 1980s, computer sharpening of ground-based images had proven that it was really an extremely compact cluster of many young stars. In passing, Hubble had made an image of this cluster, and all stars appeared separate and perfectly recognizable. And all that without any image processing! After weeks of frustration, the Hubble had delivered a success.

In almost weekly rhythm, new images were brought before the public, substantially correcting the battered image of the project. There was the great supernova of 1987 and the core region of the galaxy NGC 7454, where the local density of stars exceeded all expectations. Then came the "Einstein Cross," a spectacular gravitational lens (see page 84). A galaxy

located almost exactly in front of a much more distant quasar caused its image to split into four identical images, due to the galaxy's gravitation. Then came the first image showing the distant planet Pluto and its moon Charon as two clearly separated bodies, which had never been successfully done from the ground. The impressive first color images of the planet Saturn with its majestic rings approached the quality of those from the Voyager probes, showing that it was possible after all to do image reconstruction on Hubble images of extended objects, not only for point-like stellar objects. When a huge storm appeared on Saturn in September and produced a large white spot on its cloud cover, Hubble provided much more detailed images than any telescope on the ground.

By the middle of November, seven months (instead of three) after launch, the orbital verification of the Hubble was complete, and the science verification began. More and more time was to be dedicated to mastering the scientific instruments. The truly scientific work was to be increased, and by December 1991 the Hubble was expected to be fully tested. At the beginning of 1991 the first scientific results from Hubble data were presented.

Again there were headlines about the supernova 1987A in the Large Magellanic Cloud, which already appeared somewhat larger than a star to the Hubble. After the stellar explosion the gases expanded with a speed of 6,000 km/s. From the angular diameter of the spectacular gas ring, its inclination in space, and observations of the intensity of the explosion, its

The planet Saturn imaged with a clarity not seen since the visits by planetary probes. The image reconstruction techniques used here, requiring only a relatively modest amount of computing, opened up areas of research for the troubled satellite, which had been assumed lost (Source: NASA).

distance from Earth could be determined rather precisely as 165,000 light-years. But the spectrographs provided significant discoveries as well. They proved the existence of hydrogen clouds between galaxies in the vicinity of our own (and therefore, cosmologically speaking, in the present). Hydrogen spectral lines (the spectrographic footprints of hydrogen gas) had previously been detected only at great distances (and therefore in the youth of the universe). Only the Space Telescope found them existing today.

Extremely accurate spectra of the gaseous disk around the famous star Beta Pictoris, where planets could potentially form, were delivered by Hubble as well. Furthermore, in a spectrum of the bright star Capella, in the constellation Auriga, the spectrograph found absorption of normal as well as heavy hydrogen (deuterium), so that the ratio of hydrogen to deuterium could be determined, with an accuracy of 10 percent, to be 60,000:1. This finding contains implications of cosmological importance, since most models of the early universe called for significantly more "heavy hydrogen" in order to provide enough mass to stop the universe's expansion. Did the Hubble prove, therefore, that it would continue to expand? The results were not that clear!

The Servicing Mission

While the Hubble carried out new observations, from the solar system to the most distant objects of the universe in apparently random sequence (in reality

the telescope followed a detailed plan to utilize the available time as efficiently as possible), the first visit was being prepared at many different places. Terms like "repair" or "rescue mission" were frowned upon; NASA preferred to talk about "maintenance" or Servicing Mission 1 (SM-1). After all, routine service flights had been anticipated. Since, for cost reasons, certain components of the satellite had been designed with a life expectancy of only a few years, the shuttle flights were necessary in any case. About 200 tools and devices were prepared so that the astronauts could react to unforeseen problems. All the procedures and alternatives were practiced for hundreds of hours on mock-ups in a huge water tank. Several operations had even been tried under true space conditions on previous shuttle flights. The number of space walks, or extravehicular activities (EVAs), during the mission had by 1993 been increased from three to five; two pairs of astronauts would take turns working on the telescope. The mission had three goals: to prove that Hubble could be serviced in orbit, to increase operational reliability by exchanging degraded or failed components, and to correct the optical flaw as well as possible.

COSTAR had taken shape in only 28 months – normally, the implementation time would have been four to five years – but this time great emphasis was put on tests before the launch. This time, two completely different methods were used to test the mirrors. Fortunately, they agreed to within a hundredth of a wavelength. In addition, the entire

December 2, 1993, 4:27 A.M. local time: With seven astronauts on board, the space shuttle *Endeavor* lifts off for the first rendezvous with the Hubble Space Telescope (Source: NASA).

instrument was tested with a simulator that precisely reproduced Hubble's optical flaws. The corrector was able to combine at least 60 percent of the energy from a star into a circle with a radius of 0.1 arcsecond. But the additional mirrors absorbed about 30 percent of the light, and the image scale changed as well. The second Wide-Field Planetary Camera (WFPC2) and the cameras and spectrographs of the third, fourth, and fifth generation, to follow in 1997, 1999, 2002, and 2005, would correct the optical flaws internally. But because the corrective mirrors of WFPC2 had to be movable, the number of CCD chips had to be reduced from eight to four (see insert on page 46) for budgetary reasons.

During the three years of Hubble operations in orbit, a number of tasks had accumulated that were considered at least as important as correcting of the optics. Several primary tasks were finally established. The most important one was the replacement of the solar panels, because their shaking not only disturbed the observations but posed a threat to the satellite's structural integrity. The British company, that built the old and new solar panels, under contract from ESA, had quickly identified the fault leading to sudden deformations upon temperature changes, and convinced NASA that it was now corrected.

The mission would be considered a "minimum success" by NASA and ESA if the Hubble received either WFPC2 or COSTAR, and a "full success" if all the major tasks were completed. If not, a second mission was to follow six to twelve months later.

The servicing agenda also contained, time permitting, some secondary tasks, mostly related to electronics work. With a delay of only one day (caused by strong winds), *Endeavor* lifted off on December 2, 1993, on one of its most important missions. Astronomer John Bahcall caught the general sentiment when he called it "a question of life or death for NASA."

One of the seven crew members on *Endeavor* was ESA astronaut Claude Nicollier from Switzerland. He was given the key assignment of operating the shuttle's robot arm. Forty-eight hours after the start of the mission, it was his task to capture the Space Telescope with the remote manipulator arm and pull it into the payload bay, which he managed very successfully. On the third day, astronauts Story Musgrave and Jeff Hoffman made their first space walk, to replace a couple of gyros. Following that, they tried to retract the old solar panels of the Space Telescope, so that they could be stowed and replaced by the new ones. They took one down without any significant problems, but the second one, which had warped by up to 60 cm from its normal shape, could not be retracted. On the next day, it was Tom Akers's and Kathy Thornton's turn outside the crew cabin. They removed the warped solar panel and Thornton released it into space. Within a year it would burn up in the Earth's atmosphere. The second, rolled-up solar panel was removed and stowed in the payload bay of the shuttle. Finally, they installed two new solar panels.

A very important day of the mission was the fifth one. Musgrave and Hoffman replaced the old

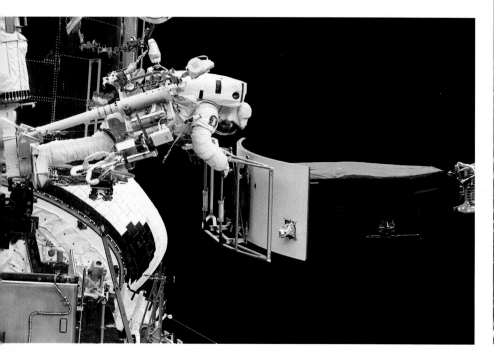

Wide-Field Planetary Camera with the new one – this already guaranteed the "minimum success" of the mission. On the sixth day, after Akers and Thornton installed COSTAR and a memory extension for the on-board computer during their second EVA, NASA and astronomers all over the world were jubilant. On the seventh day, Musgrave and Hoffman replaced one of the drive electronics boxes for the solar arrays. Because these units had not been designed for

Hubble is repaired, the new solar panels are unfurled, and redeployment is imminent (Source: NASA).

of one of the magnetometers, discovered during the space walks, was fixed using mylar foil from the old Wide-Field Planetary Camera. The canisters with the new solar panels would not deploy right away, but some pressure applied by hand helped. The astronauts had spent a total of 35½ working hours in space. They left the telescope to itself on December 10 and returned triumphantly to Earth. Three days later, on December 13, 1993, at exactly 12:26 A.M. local time, the space shuttle touched down at Cape Canaveral. The total cost of the servicing mission, including the flight of the shuttle, was $674 million, of which $100 million was a direct result of the optics flaw. But because Perkin-Elmer had never been convicted of gross negligence, and also because NASA had obviously not fulfilled its supervisory duties, only a fraction of this amount was recovered.

The necessary adjustments to all the new mirrors went off without any problems. By the beginning of 1994, rumors were trickling out that the mirror adjustments were proceeding faster than expected and that the first sharp images were available. A large press conference was announced for January 13, 1994, again at the Goddard Space Flight Center where the optical problems had been made public three and a half years before. The constitution of the panel alone gave away the fact that there was cause for celebration: NASA administrator Dan Goldin was there, as was the science adviser to the president. History was made when the energetic Senator Barbara Mikulski stood up, produced two

replacement in orbit, the exchange took considerable time. On the following day the Hubble was moved out of the payload bay and released again into space.

The astronauts had completed all their tasks by the end of the fifth space walk! There were only very few small problems – once one of the equipment doors on the telescope had to be forced closed; once a little screw got away and had to be chased through the entire payload bay. Some damage to the insulation

Left: Hubble shortly after deployment, with all repairs completed. Slowly *Endeavor* moves away . . . and then the Space Telescope appears to the exhausted astronauts as only a bright star between the crescent moon and the glow of Earth's atmosphere (right) (Source: NASA).

Far left: One of the best ground-based images of the "Grand Design" spiral galaxy M100, taken with the 4-m (160") Anglo-Australian Telescope. Center: An image of the galaxy taken with WFPC-2, after the repair. Top right: An enlarged cutout of the same image. Bottom right: The identical area in comparison, from an image taken for that purpose with WFPC-1, shortly before its removal (Sources: ASP, J. Trauger/JPL, and NASA).

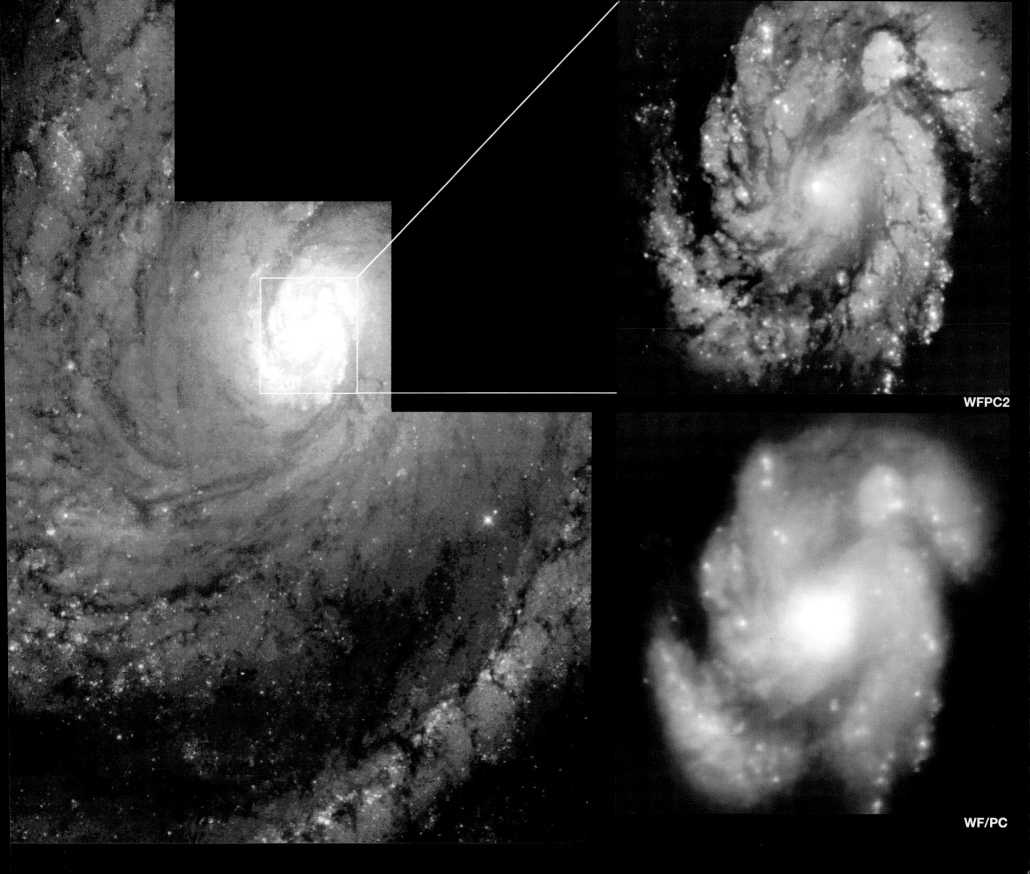

WFPC2

WF/PC

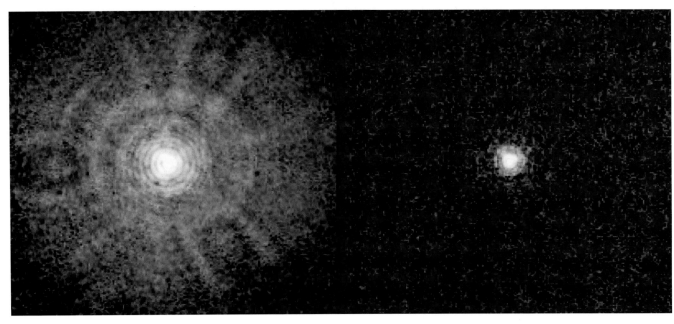

"The trouble with Hubble is over!" The left image shows the image distortion, distributing large fractions of the light over a halo of an arcsecond diameter – on the right a comparable image after the servicing mission (Source: ST ScI).

images, and announced: "I am delighted to be able to announce today, after the launch in 1990 and the early disappointments, *the trouble with Hubble is over!*" Her first picture showed images of single stars, highly magnified, before and after the installation of COSTAR. The hideous halo of light, with its circles and tentacles, had disappeared. Even more impressive was Mikulski's second picture, an extremely sharp color image of part of the spiral galaxy M100. The comparison with an image of the same galaxy, taken shortly before the servicing mission, dramatically proved its huge success. "I believe these images are tangible proof that not only has Hubble been repaired," concluded Mikulski, "but that NASA is

on its way to repair the attitude that led to these problems." On pages 58–59 we show before-and-after images of the spiral galaxy M100, alongside one of the best ground-based images of this spectacular galaxy. These photos impressively show the value of the repaired Space Telescope and the usefulness of optical telescopes in space.

Hubble had survived its black eye and could now tackle all the tasks that had had to be withheld, in particular the exploration of the most distant realms of the universe. Let us follow it on its journey from the most distant objects of the universe, through neighboring galaxies, stars, and nebulae in our own galaxy, to the planets of our solar system.

Part 2

The New Window to the Universe: The Hubble and Astronomy

The Great Questions: How Large and How Old Is the Universe?

Our universe is filled with stars, some of which may have a planetary system like our sun's – or may not. Some of these planets may harbor life – or may not.

Large accumulations of stars are called galaxies, and our galaxy, in which the sun moves in its orbit, is called the Milky Way because it looks like a faint white path or river in the night sky. Galaxies are classified according to their appearance: spirals, barred spirals, and elliptical galaxies. But for astronomers this classification is only a first step toward deeper understanding, a key to galactic structure and evolution. The longer we study them, the more they turn out to be mysterious objects containing answers to deep cosmological questions.

One of the great unsolved questions of astronomy is the age of the universe. Edwin Hubble first showed us that its age is not infinite, and the Hubble Space Telescope could be the tool that helps us to definitively answer this question.

Edwin Hubble and his work started us on the road to understanding the true dimensions of the universe. His discovery of pulsating stars in the Andromeda galaxy provided the opportunity to measure the distance of a galaxy without difficulty, which he did starting in 1923. After determining a good number of distances of such galaxies, Hubble found in 1929 a linear relationship between the distance and the redshift in the spectra of galaxies – the Hubble Law.

Redshift means that the light arrives at the observer with a longer wavelength than it was emitted with, so that, for instance, blue light can be seen as green or even red. A shift of the spectral lines toward the long-wavelength, red end of the spectrum occurs when the light source moves away from the observer – in physics, this is known as the Doppler effect. Since the speed of light cannot change, when the light source moves away from the observer the distance between wave peaks increases. In other words, the wavelength gets longer. But in the universe the cosmological rather than the simple Doppler operates. The entire universe expands, and with it the light waves moving through it. The more distant a galaxy is from us, the longer its light takes to arrive, and the more it is "redshifted" when it arrives on Earth. The degree of redshift can be used as a measure of a galaxy's speed relative to Earth.

The relation between distance and redshift has the surprisingly simple form $v = H_0 d$, where v is the speed expressed in kilometers per second, d is the distance in megaparsecs (1 megaparsec = 3,260,000 light years), and H_0 is the Hubble constant. The Hubble constant is a very important number: it measures the speed of expansion of the universe. A helpful way to understand Hubble's law is to realize that it is just stating that the velocity a galaxy has, and the distance it is from us, are proportional. *This would be true if all the galaxies started moving at one unique instant in the past.* Those moving twice as fast are twice as far away, those moving three times as fast are three times as far away, and so on. This evidence that everything started moving at the same time leads directly to the idea of a "Big Bang" at

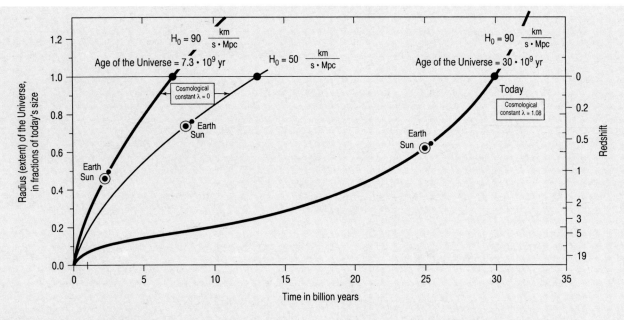

Models of the Universe

The graphical representation above, after Professor W. Priester, attempts to illustrate the variety of possible models for the universe in a single picture, providing the reader with a vehicle to get a feeling for the dimensions of modern cosmology. The horizontal axis is time, measured in billions of years since the Big Bang on the left. The vertical axis on the left shows the radius, or extent, of the universe in fractions of today's radius; on the right vertical axis the equivalent redshift is given, which a celestial object would exhibit today, as we see it at an earlier point in time, defined by the world radius at that time. The relation between redshift and world radius is fixed, but the relation between world radius and time is not: Depending on the adopted values for the Hubble constant (describing the rate of expansion of the universe, H_0 indicates today's value) and the cosmological constant, the history of the universe takes different paths. The graph also in-

dicates the times when the sun and earth were formed.

Left: $H_0 = 90$ km/s/Mpc (as indicated by some of Hubble's measurements); cosmological constant $\lambda = 0$. The current universe is hardly 8 billion years old, which is in direct contradiction to the age of the oldest stars in the Milky Way – they are definitely twice that age.

Center: $H_0 = 50$ km/s/Mpc; cosmological constant $\lambda = 0$. In this model, the universe is about twice as old, and the contradiction to the age of the oldest stars vanishes, but such a small Hubble constant would indicate that some Space Telescope results contain significant errors. Because of its mathematical simplicity ($\lambda = 0$), this model continues to have supporters.

Right: $H_0 = 90$ km/s/Mpc; cosmological constant $\lambda > 0$. This model is gaining support, even if it introduces the "complication" of a non-zero cosmological constant. Now the age of the universe is significantly higher (about 30 billion years), since the expansion rate of the universe was very low for several billions of years, and only picked up speed more recently (that quiescent phase could also be helpful in explaining the formation of the first galaxies).

It is still too early to decide definitively between these models of the universe, but the curves can at least be used to determine the distance to a remote object (a galaxy or a quasar) of a given redshift, depending on the model (and therefore indicating the uncertainty). From the point representing today on the preferred curve, a vertical line is drawn down to the corresponding value for the observed redshift; the distance from that point to the intersection with the respective curve gives the distance of the object in billions of light years. It should be noted that the distances up to a redshift of about 1 differ very little for the more popular models, while for higher redshifts the distances are much greater for the model on the right.

the start of the expansion, and tells us that the phrase "age of the universe" has a real meaning – it is the time since the universe started expanding.

Whereas Hubble derived a value of 530 kilometers per second per megaparsec (km/s/Mpc) for his constant in 1929, today's more accurate values fall in the range from 50 to 100 km/s/Mpc, a change of almost 10 times. One can see what that means: The larger the Hubble constant, the faster the universe expands, so the younger it is. Conversely, if the Hubble constant is lower, the expansion is slower and must have taken a longer time. The inverse of the Hubble constant must be a measure of the age of the universe. Suppose somebody has made a movie of the entire life of the universe. If we run the movie backward, all the galaxies now moving away from us would begin converging instead. After a certain time, there would be a collision, and this collision in the reversed movie would be the Big Bang – the birth of the universe in a hot, dense explosion. This reasoning makes apparent the importance of the redshift in cosmology. The receding of galaxies can be explained by a common origin of this motion, the Big Bang – the now generally accepted model of the origin and evolution of the universe.

Aside from the Hubble constant, there are other quantities describing the structure of the universe that astronomers seek to measure: the acceleration parameter, the density parameter, pressure, a possible cosmological constant, and the curvature of space are all related to each other in subtle ways. The acceleration parameter indicates how much the expansion of the universe has slowed down since the Big Bang. We know that the expansion is gradually slowing due to the gravitational pull of all the matter in the universe on all the other matter. The density parameter measures the amount of matter in the universe, and indicates whether its density is high enough that the gravitational pull between the galaxies is large enough to eventually halt the expansion of the universe, and cause it to collapse. The density necessary to halt the expansion is called critical density. The pressure is composed of radiation pressure and the motion of the galaxies – in general, it can be neglected. The cosmological constant indicates whether or not there might exist a repelling force, proportional to distance, acting between faraway objects, something like the opposite of gravity. (There is currently no good evidence for this.) Einstein taught us that the pull of gravity can be strong enough to curve space itself, and this curvature can be described by a number that ranges between -1 and $+1$. If the curvature is 0, then space is flat (or Euclidean); if it is $+1$, then space is positively curved, sort of like a sphere, but through a higher dimension – that means it doesn't go on forever, and won't keep expanding forever. If the curvature is -1, then space is hyperbolically, negatively curved, sort of like a saddle that goes on forever in all directions. In this case, the expansion of the universe would never stop.

Today, the theory of the inflationary universe is widely discussed. In its simplest form it states that there is no additional repelling force in today's universe (i.e., the cosmological constant is zero), that space is flat, meaning that parallel lines never meet and the angles of all triangles add up to exactly 180°, and that the density of the universe is equal to the critical density. (That is not true in curved universes. For instance, in a spherically curved universe, parallel lines don't remain parallel but meet in the far distance, just as lines of latitude start out parallel at earth's equator but meet at the poles.) However, all the observable material in our universe, all the galaxies, stars, planets, etc., adds up to only about 1 percent of the critical density, so that it has to be assumed that if the universe is flat and won't go on expanding forever, a large fraction of matter is not luminous. Only a very small part of this "dark matter" can be explained by nonradiating gas, dark planets, or black holes. It is conjectured that it might consist of massive elementary particles that have only weak interaction with "normal matter" and have not yet been detected. Many open cosmological questions remain. Regardless of the cosmological model, the equations yield values for the age of the universe between $2/3 \times (1/H_0)$ and $(1/H_0)$. The former (lower) value corresponds to the critical density, and the latter to a density approaching the observed density of normal matter. For $H_0 = 100$ the age of the universe lies between 7 and 10 billion years, and for a value of $H_0 = 50$, about twice that. The oldest observable stars are thought to have an age of about 15 billion years. Since the oldest stars in the universe can't be older than the universe itself, these two different astronomical studies disagree unless the smaller value of the Hubble constant is correct. However, calculations of the lifetimes of stars are based on computer models, and either stars or the universe may be more complicated than we know. Nevertheless, the best models for the ages of stars in globular clusters show that the clusters are between 11 and 21 billion years old, indicating a Hubble constant of at most 60, if we do not want to allow stars to be older than the universe. All these theoretical considerations suggest how important it is to get better distance determinations of galaxies, based on a multitude of well-calibrated objects. These key objects include pulsating and exploding stars in galaxies – so-called standard candles – whose intrinsic brightness is known, and whose apparent faintness therefore indicates distance. Only these can provide a precise answer for the age of the universe, and such measurements are possible only with a space telescope like Hubble.

Astronomers are making these measurements in ways Edwin Hubble would have recognized. First, reliable distances to galaxies are determined, then redshift is measured to indicate velocity, and finally the Hubble constant is derived, yielding at last the age of the universe. The process has some interesting details.

One of the main tasks of the Space Telescope is to find pulsating stars in distant galaxies. These are called Cepheid variable stars, and once the period of a

small individual random motions that can confuse our measurements. But since the latter are small and the former increase with distance, cosmological redshift measurements become more accurate with increasing distance. Aside from pulsating stars, globular clusters and supernovae are used as "standard candles" in distance determinations as well, as we will discuss later on.

Cepheids

Cepheids are giant stars, a late developmental stage of massive stars, that have moved away from the main sequence of hydrogen-burning stars and entered the so-called "instability strip." Certain conditions in their outer layers cause these stars to begin to pulsate. At the beginning of this century, Henrietta Leavitt (1868–1921) and Ejnar Hertzsprung (1873–1967) recognized a correlation between the period of the pulsation and the luminosity or absolute magnitude of these stars: the longer the period, the more luminous is the Cepheid variable. We know hundreds of Cepheids in our own Milky Way, including many whose distances and luminosities are known with enough accuracy to let us calibrate the period-luminosity relation. Once the period of a Cepheid in a distant galaxy is known, a comparison of the luminosity predicted by the period-luminosity relation with the apparent magnitude yields the distance of this pulsating star.

The three enlargements show a Cepheid in the spiral galaxy M100, with clearly visible variations

A Cepheid variable, pulsating in the galaxy M100: After its repair, Hubble was also able to see this type of star in galaxies in the Virgo cluster and to follow its change in intensity. The three inserts show blowups around a Cepheid of changing brightness (Source: W. Freedman and NASA).

pulsating star of this type is known, its actual or intrinsic luminosity can be derived, since a precise relation exists between period and luminosity. In comparing the luminosity with the apparent brightness, the distance to that star can be determined – and thus the distance of the galaxy. Its redshift can be measured by a simple spectroscopic observation with a ground-based telescope. Once we know the distances and redshifts of many galaxies, we can derive the Hubble constant much more accurately than before.

In addition to the "cosmological" redshifts due to the expansion of the universe, however, galaxies have

in brightness. The brightness of this faint Cepheid, which could not be observed as a single star by any ground-based telescope, varies with a period of 51.3 days. From that the accurate distance of this galaxy can be derived: about 56 million light years.

M100 is part of the Virgo cluster, a well-known galaxy cluster in the constellation Virgo. It contains many large and small galaxies concentrated around at least two centers. The spatial extent of the entire cluster, and the motions of the individual galaxies with respect to each other under the influence of their gravitation, are very complicated. The receding velocity of the Virgo cluster is therefore not simply the "Hubble flow," as the cosmological expansion is called. However, once the distance of the Virgo cluster is known accurately, then several other secondary distance indicators can be calibrated and applied to more distant galaxies, whose motions must show just the Hubble flow. Again, this extension of observations to greater distance should increase the accuracy of our measurement of the expansion rate and age of the universe. So the measurement of the distance of M100 was only a first step.

These observations, carried out by American astronomer Wendy Freedman, seem to indicate a Hubble constant of about 80 km/s/Mpc. However, there are probably some individual peculiarities of stars that could cause small variations from the period-luminosity relation, and certainly astronomers' measurements are not perfectly precise. So to arrive at a more precise result, many Cepheids

The elliptical galaxy NGC 4881 in the Coma cluster and its vicinity. The faint points of light surrounding the galaxy are globular clusters, and their brightness distribution provides indirect evidence for the distance of the galaxy. The surprising result: the upper limit for the Hubble constant derived from these data is lower than the high value obtained from distance measurements of M100 using Cepheids (Source: WFPC-2 Team and NASA).

should be observed and the results averaged. (This is like any survey that involves sampling: too small a sample leads to uncertain results.) As more galaxies of the Virgo cluster are measured with this method, the accuracy of the actual value of the Hubble constant increases. But the current trend of the observations is clear: Cepheids in several other Virgo cluster galaxies and more of these pulsating stars in M100 itself have been detected – and large values for the Hubble constant resulted in all cases. This would mean that the universe is much younger than previously thought. Is it necessary then to rewrite the complete theory of evolution of the universe?

Globular Clusters

It is not that simple. The image here shows a field of distant galaxies. It is assembled from 16 exposures of 15 minutes each. The brightest object in the field is the elliptical galaxy NGC 4881, situated in the outer parts of the Coma cluster. Its redshift corresponds to an expansion velocity of 7,000 km/s. Next to this galaxy of 13th magnitude one can see a fainter spiral galaxy of 16th magnitude, which is also part of the Coma cluster, as well as several foreground stars of our own Milky Way. All other galaxies in this image lie far beyond the Coma cluster in the depths of the universe; one object shows the merging of two galaxies.

In the vicinity of NGC 4881 we can distinguish faint, point-like objects, which are globular clusters of this galaxy, and can thus be used to determine its distance. The globular clusters of our own Milky Way have an absolute magnitude of -7.6^M, and in other galaxies in our vicinity the majority of them appear in the range from -7^M to -8^M. This maximum cannot be seen in this image of NGC 4881, even though it shows objects down to an apparent magnitude of 27.6. From this it can be determined that the Coma cluster is at a distance of more than 100 Mpc, and that the value of the Hubble constant is under 70 km/s/Mpc, significantly different from the 80 km/s/Mpc derived by Wendy Freedman using the Cepheid method. To confuse the picture even more, however, yet another distance determination of the Coma cluster, which relied on comparisons of the brightness distribution in the nearby galaxy M87 with those in the Coma cluster, yielded a high value of the Hubble constant, 78 ± 11 km/s/Mpc. The distance to the Coma cluster is an important factor for the cosmological distance scale, since at that distance the random motions of galaxies cause only negligible deviations from the cosmic expansion.

The Hubble carries its name not least because of the expectation that the telescope could help determine the Hubble constant with an accuracy of 10 percent after a few years of observations. In December 1995, at a conference in Paris, a major debate was staged between astronomers who have found a high Hubble constant using the Hubble telescope, and those who have found a low constant using the same equipment. While both factions left the stage unshaken, they could at least agree on one thing: the Virgo cluster, because of its enormous size and very complex inner structure, is a very treacherous rung of the cosmological distance ladder. Wendy Freedman vowed that Hubble would keep its promise, and by 1998 we would know the value of the Hubble constant to within 10 percent. Formally speaking, this goal is indeed being approached by the Virgo-Cepheid method. But we will be certain only if several independent methods yield the same answer. Nevertheless, the Hubble can help improve several of the techniques already applied from ground-based telescopes. One of the most popular is the utilization of supernovae in distant galaxies as standard candles of known brightness.

Supernovae are stellar explosions, which will be discussed in detail later. For now, it is important only that a particular class of them always appears to reach a very similar maximum intrinsic brightness. But it had not been proven that these supernovae truly have a fixed maximum brightness, nor was it possible until recently to calibrate this brightness well. Here the Hubble can help. Using the Cepheid technique, the distances to two nearby galaxies were determined. These galaxies were too close to exhibit significant motion from cosmic expansion, but they contained two type Ia supernovae, whose apparent maximum brightness had been measured. As these are compared with supernovae in much more distant galaxies, for which no direct method of distance determination is available, the distance to the more distant galaxies can be determined. This technique permits yet another estimate of the Hubble constant.

The value from this method appears to be close to 50 km/s/Mpc, very different from the determinations described above, indicating a much older universe.

So what is Hubble's contribution to the determination of the Hubble constant? What is its value, and what is the age of the universe? In 1995, it is still too early for a final result, although most results seem to favor a somewhat younger universe than previously thought. If the Hubble constant is indeed closer to 100 than to 50 km/s/Mpc, then at least one conclusion is inescapable: The simplest cosmological model, with a cosmological constant of zero (meaning there is no repellent force), cannot be correct, because it appears that the oldest known stars are more than 15 billion years old. Either that, or our models of the interior structure and evolution of normal stars, which are thought to be one of the better-understood areas of astrophysics, have a major problem. It would be an exaggeration to state that the theory of the Big Bang is in jeopardy, since many of its predictions have turned out to be true, but the details of how the expansion has occurred remain uncertain. A positive cosmological constant could reconcile a large Hubble constant with a higher age of the universe, since such a value for the cosmological constant leads to a long period of very low expansion during the early phases of the universe. It has even been suggested that a "rest phase" of about a billion years in the cosmic expansion could contribute significantly to the formation of galaxies (see also the diagram on page 64).

So it is still unclear whether we live in a relatively simple universe without a cosmological constant and

The UV spectrum of the distant quasar Q0302-003. The cutoff of the spectrum at 130 nm is a clear indication of the existence of intergalactic helium. This was Hubble's first major discovery after the servicing mission (Source: ESA).

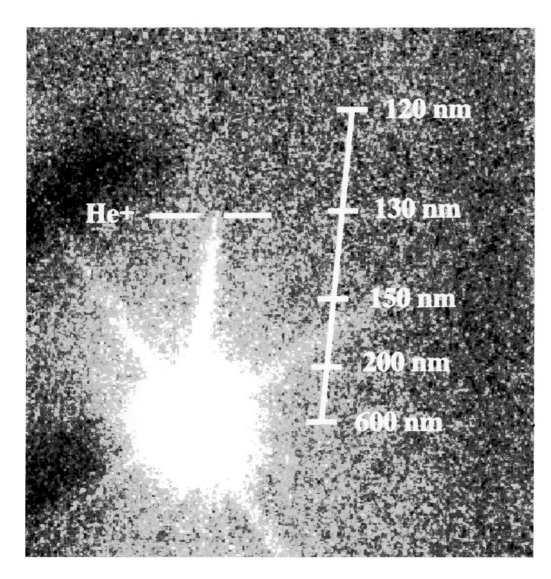

He+

120 nm
130 nm
150 nm
200 nm
600 nm

with low expansion, or in a more complex universe with a higher Hubble constant. But the differences between the various world models have very concrete consequences. For very distant celestial objects we can only make vague statements about their distance, even if the redshift is known and a value for the Hubble constant is agreed upon. As the diagram on page 64 shows, the distance to an object of given redshift varies greatly with the world model. *All* statements about distances of the order of billions of light years are necessarily inexact. So we can still hope that Hubble will contribute decisively to the fundamental questions of distances and age of the universe. But for now the great riddles remain unsolved.

The Hubble and the Big Bang Helium

Another fundamental question in the field of cosmology revolves around the formation of the universe. The most widely (but not universally) accepted theory is the Big Bang. The Hubble found another piece of evidence supporting this theory.

The Big Bang theory predicts that the simplest element, hydrogen, should have been formed in great quantities during the cooling of the hot fireball of the primordial explosion. However, about every tenth atom would become helium. About every ten-billionth atom should become lithium, and no other elements beyond these first three should form. (Other atoms are produced inside stars and scattered into space by supernovae.) Most of the helium gas would be used up in the formation of galaxies and the first generation of stars, but a few traces should have remained in the spaces between the galaxies. Hubble seems to have discovered evidence for the existence of such helium.

The adjacent picture shows a spectrum taken with Hubble's Faint-Object Camera. With a prism, the light of a distant quasar has been dispersed into its component colors, arranged vertically here, and marked with a wavelength scale. Most of the light of this quasar (Q0302-003) reaches us in the red part of the spectrum (the bright spot on the bottom), due to the large redshift of this distant object. However, the ultraviolet range of the spectrum was of special interest, as it cannot be observed from the ground. It has been dispersed upward into a near-vertical spike of light – and particularly exciting is the fact that this spike disappears abruptly at a wavelength of 130 nm. At all wavelengths shorter than 130 nm the helium between the distant source and us absorbs the light, and the spectrum is chopped off. Is this proof that clouds of helium exist in intergalactic space?

A telescope reaching even further into the ultraviolet was required, so that the helium absorption could be detected for quasars with lower redshifts. The Hopkins Ultraviolet Telescope (HUT) was included in the Astro-2 mission aboard the space shuttle at the beginning of 1995, and several hours of observing time with its spectrographs were dedicated to the brightest known quasars. This telescope was specifically built for the detection of intergalactic helium, and in June 1995 the HUT astronomers were able to present their results: The telescope had unambiguously measured

this effect on other quasars as well, in addition to the original Hubble detection. Another piece of the puzzle of Big Bang theory had been found! The adjacent picture by itself may be one of the least impressive ones taken by Hubble, but from the point of view of astrophysics, it may be one of the most important images. Similar Hubble spectra have played a major role in showing that the fourth and fifth elements, beryllium and boron, are *not* seen in the very oldest stars, again confirming the Big Bang prediction that only the first three elements are primordial.

In 1995, Hubble made yet another important contribution to the exploration of the not-so-empty space between galaxies. The Space Telescope was able to show that many narrow absorption lines, which had long been known to exist in the spectra of quasars, were not caused by intergalactic clouds but by gigantic gas halos around normal galaxies. Only Hubble, with its high sensitivity in the ultraviolet, can observe these absorption lines also at lower redshifts, allowing astronomers to see these galactic features for the first time. They got a big surprise: For about half of the absorption lines the responsible galaxies were detected, but they were not always close to the line of sight. It turned out that totally normal galaxies are surrounded by extended hydrogen clouds, reaching a diameter of up to 15 times the optical image of the galaxy! The origin of this gas is unclear. According to all theories about galactic systems, gas could not be maintained at those distances.

Island Worlds in Space and Time: Galaxies and Quasars

The large size of the universe, combined with the finite speed of light, means that every glance into the depths of the universe is also a glance into the past. What is impossible for humans is no problem for the Hubble: to bring the past to life. However, very distant and therefore very old objects appear very small. The limited resolution of ground-based telescopes had always clouded our view of the distant past. Many questions have not yet been answered: How do galaxies evolve over billions of years? Is there a transition from spiral galaxies to elliptical galaxies, or the other way around? Does the number of galaxies decrease over time because of "cannibalism," galaxies merging, or big galaxies pulling in little ones? Do new galaxies condense out of hitherto "unused" hydrogen and helium in the universe? Hubble is in the process of providing important information to clarify these questions.

Our first image (on page 74, left) shows an area in the constellation Sculptor, about the size of the bowl of the Big Dipper, taken with the United Kingdom's Schmidt telescope in Australia. In its center there is a faint cluster of galaxies, invisible in this image. All of the stars are in the foreground in our own Milky Way. The middle image comes from a 4.7-hour exposure with the Hubble, showing objects down to magnitude 28.5. The bright star-like object in the center is the quasar Q0000-263. In its vicinity, there is a cluster of galaxies, consisting of fourteen objects, probably associated with the quasar. All of these objects are at a distance of more than 10 billion light years and demonstrate the appearance of the universe only about 2 billion years after the Big Bang. In a sense, Hubble has taken a picture of our own nursery. This image, together with the following one, represents one of the deepest glances into space, and therefore into the past—impossible for humanity up to now.

The left image on page 75 shows one of the longest exposures of the Space Telescope. This 18-hour exposure of a field in the constellation Serpens (the snake) is constructed from several individual images taken between May 11 and June 16, 1994, and shows objects down to magnitude 29. Many galaxies with distances from 5 to perhaps 12 billion light years can be seen in this field. Some of these objects are at a distance of 9 billion light years, among them the unusual radio galaxy 3C324 in the center of the image and in the enlargement on the bottom right.

The high resolution of the Space Telescope makes it possible to see the details that allow astronomers to classify these galaxies into spirals and ellipticals. It turns out that in earlier times galaxy clusters contained significantly more spiral galaxies than today. Several of these galaxies show strange structures, and we even find irregular fragments of galaxies.

Only a few of these galaxies appear to be normal spirals, but there are several deformed and merging objects. Such a group can be seen in the upper right image. In the early universe there appears to have been a large likelihood of close encounters of galaxies, and thus of galaxy mergers.

The cluster in Serpens also contains a number of red galaxies, similar to today's elliptical galaxies.

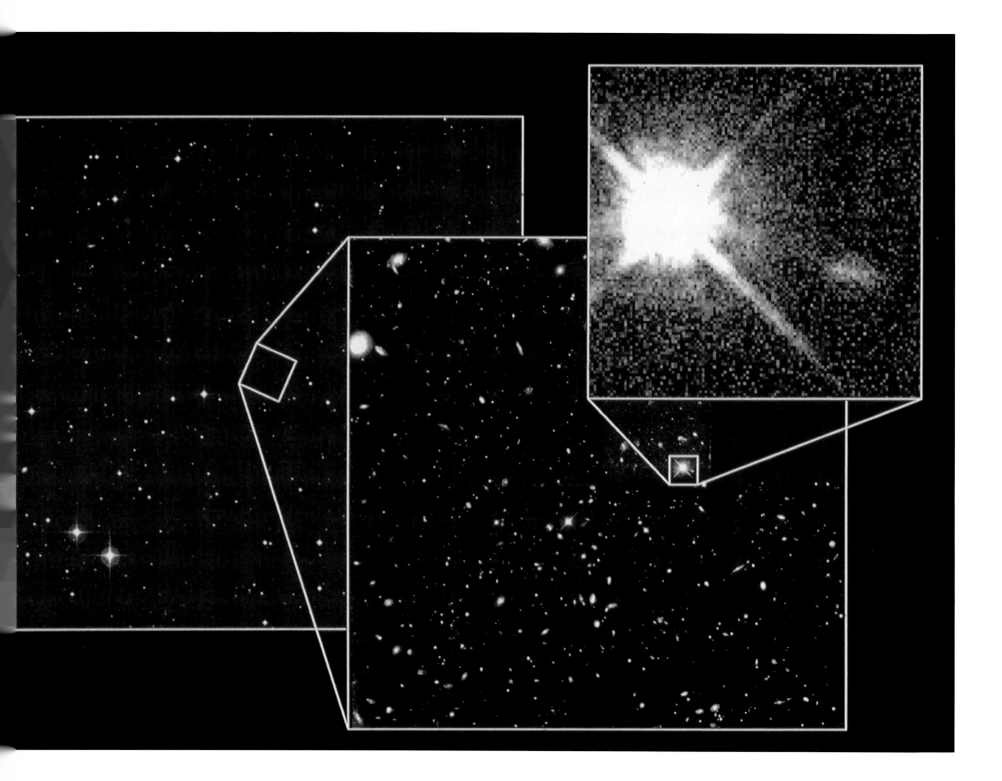

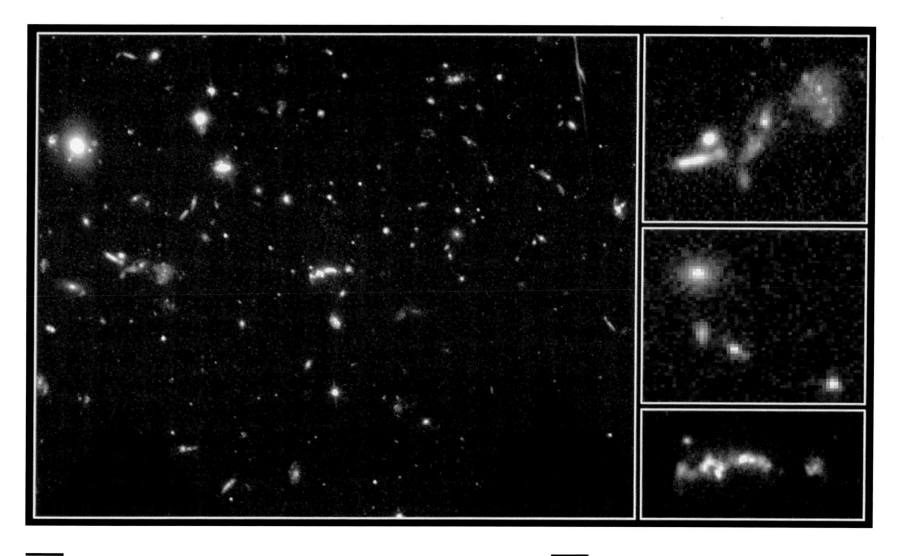

Left page: A view of our childhood, the young universe. This is one of the most distant clusters of galaxies known, at a distance of about twelve billion light years, next to the quasar Q0000-263 (Source: D. Macchetto, M. Giavalisco, and NASA).

A deep glimpse into the past. In the center of the large image a radio galaxy and a faint cluster of galaxies, probably in its vicinity, are visible at a distance of about nine billion light years. Images like these are proof that the universe looked different at different times. Therefore it undergoes an evolution and must have had a beginning (Source: M. Dickinson and NASA).

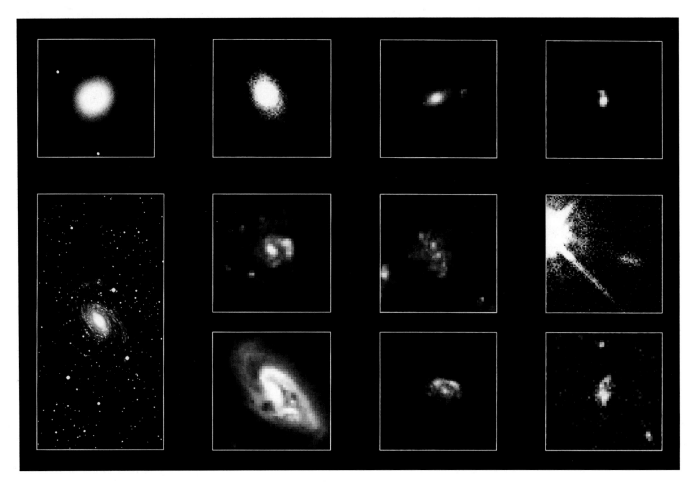

The evolution of galaxies over billions of years (Source: A. Dressler et al. and NASA).

They consist of stars that formed very soon after the Big Bang. Such a group of elliptical galaxies can be seen in the center right image on page 75. Elliptical galaxies, and the stars that are contained in them, appear to have formed in the early universe and to have maintained their structure without any significant changes.

All these observations make it possible to illustrate the structure of elliptical and spiral galaxies over the life of the universe. This has been attempted in the sequence of images above. Even Hubble cannot overcome the fact that young objects are visible only in the extreme depths of space, that they appear very small, and that they therefore can be observed only with much lower clarity than galaxies in our immediate vicinity. But it is still possible to distinguish different classes of galaxies in the distant past. The left column shows the two prototypes that can be found in the present universe, about 15 billion years after the Big Bang: elliptical galaxies, consisting of old stars (top), and spiral galaxies, where even today stars are being formed in their disks of gas and dust (bottom). (Note that since ages and distances scale with the Hubble constant, all the distances to these galaxies are as uncertain as the constant itself. So 15 billion years is very uncertain.)

Looking at the top row, we can see that elliptical galaxies have hardly changed their appearance. Even the youngest objects on the right side show the brightness profile characteristic of elliptical galaxies.

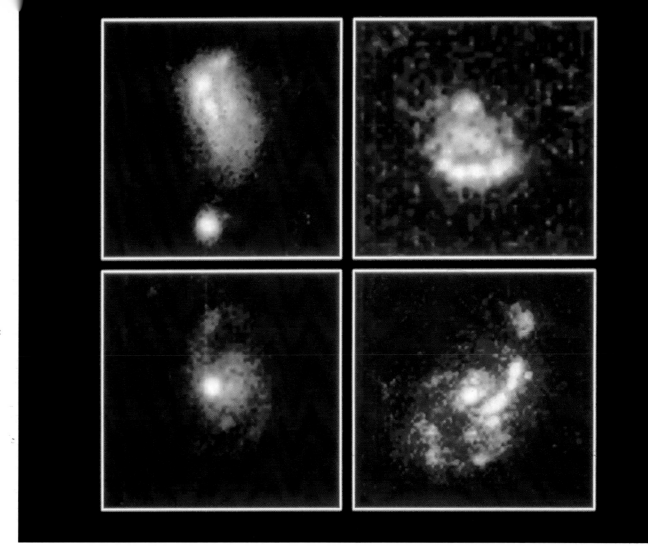

Young galaxies in color (see also page 78), as Hubble is able to see them at different epochs of the universe (Source: NASA).

In contrast, spiral galaxies, which generally appear symmetrical today, had much more irregular shapes in earlier times (center left), including nonuniform, extended star-forming regions (starbursts; center right). In the very early universe these differences may not be that pronounced; however, elliptical and spiral galaxies can still be distinguished in spite of the great distance (extreme right). In 1995, several teams jointly reached another surprising conclusion: the farther back we move in time, the greater the number of irregular or outright strange galaxies we see, even though the number of spirals and ellipticals stays about the same as today. The fact that the universe is evolving has never been so conspicuous. Piecing together a complete picture of the evolution of galaxies, however, is a major task that is just now being addressed.

Galaxy interactions and collisions occurred not only in the early universe but also in the "recent past," at lower redshifts and with objects appearing at larger angular size, as can be seen in the spectacular image of a frontal collision of two galaxies. This so-called Cartwheel Galaxy, in the constellation Sculptor, is at a distance of 500 million light years.

The ring-shaped structure appears to be a direct result of the intrusion of a smaller galaxy – probably one of the two objects to the right of the ring – into the central part of the original galaxy. This collision gave rise to a shock front in the disk, consisting mainly of gas and dust, which traveled through

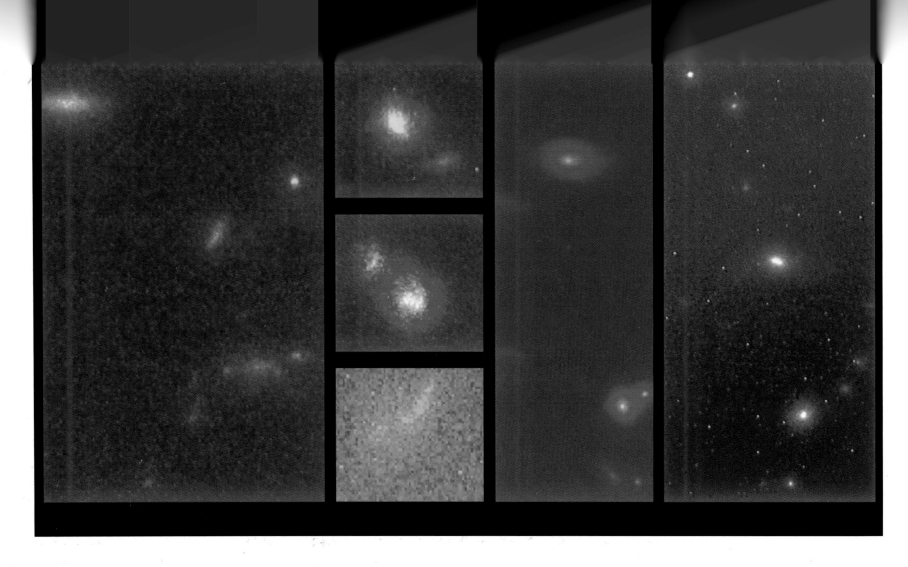

the disk like a tidal wave, with a velocity of over 300,000 kilometers per second, causing an intense period of star formation. In the image we can see bright blue nodules; these are clusters of massive young hot stars. We also see extended rings and bubbles caused by supernova explosions, the final stages of the development of such massive stars. The ring has a diameter of 150,000 light years; the Milky Way would fit comfortably inside. Despite this size difference it appears that this galaxy had similar characteristics to our own, and that it was modified by the collision but is now in the process of regaining its original structure. The first indications of spiral arms ("spokes") are visible between the central "eye" and the outer ring. It is still unclear which one of

the two neighboring galaxies was involved in the collision. The red galaxy contains no gas, which would have been swept away if it took part in the collision. Astronomers expect, however, that it takes a while for such a partner in a collision to evolve into a red galaxy. The blue galaxy, on the other hand, shows traces of interactions and current star formation – in contradiction to older theories, but perhaps explainable if it has recently emerged from a collision.

Collisions are not rare in the depths of space. They can occur more frequently in the life of a galaxy than in the life of a motorist, and can dramatically influence a galaxy's characteristics. It may obtain a new shape or even be fully integrated into its partner, cannibalized in the collision. The total number of galaxies should

Top: Hubble's yield from the search for young galaxies.

Facing page: The "cartwheel" galaxy, at a distance of 500 million light years in the constellation Sculptor, resulted from a collision of a smaller galaxy with a normal spiral galaxy (Source: K. Borne and NASA).

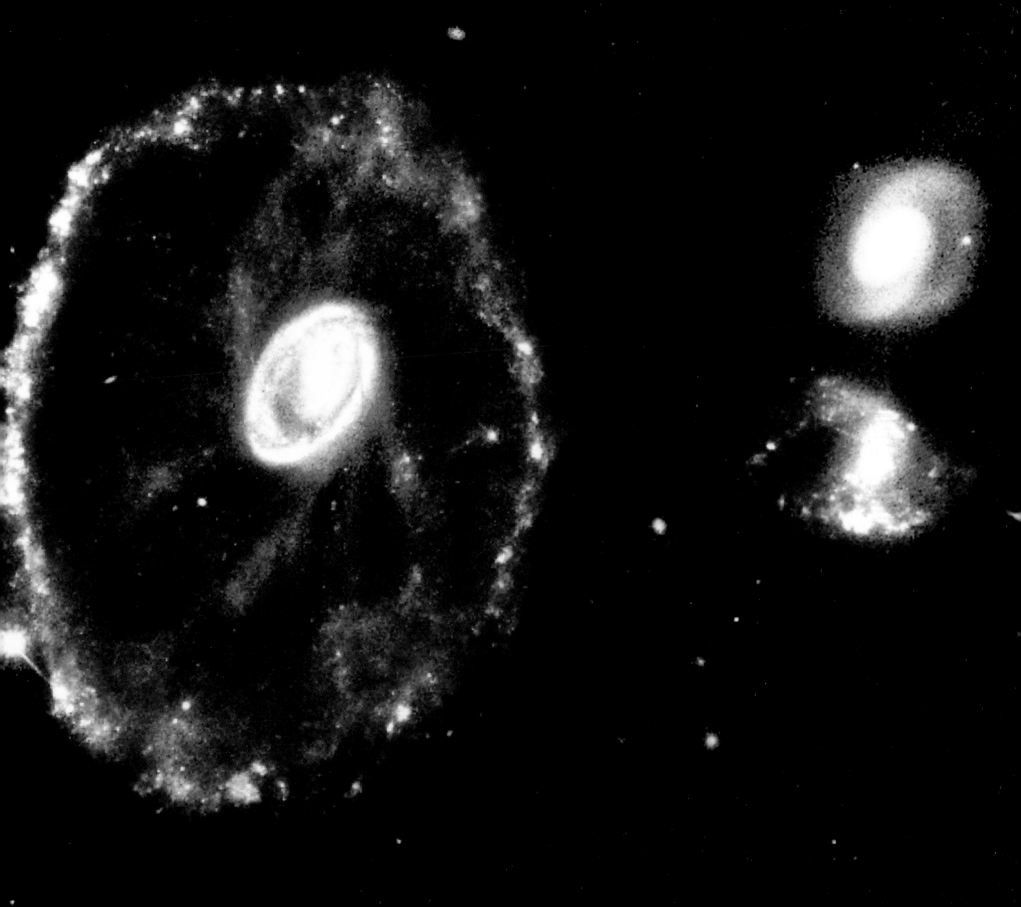

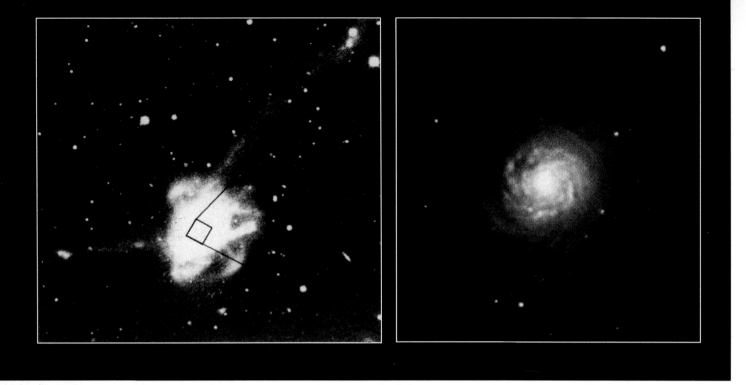

therefore decrease from the early times to today, as long as no new galaxies are formed from the remaining materials in space.

Some astronomers believe elliptical galaxies are formed by collisions merging two or more spiral galaxies, but one problem remains: spiral galaxies are accompanied by relatively few globular clusters – compact accumulations of the order of a hundred thousand stars – whereas elliptical galaxies show many clusters. Where should they have come from in such a merger or fusion scenario? Hubble has contributed important data to this controversy. Its observations of a spectacular collision of galaxies (the resulting object is called NGC 7252) appeared to provide the answer: at the place where the two spiral galaxies had encountered each other, several dozen young globular clusters had formed; their formation appeared to be an unavoidable by-product of galaxy mergers. Opponents of this theory, who have investigated over 600 young star clusters as a result of another collision, dispute this interpretation. The new accumulations of stars may not be globular clusters but so-called open clusters, where stars are much more distant from each other, and which differ

in size from a typical globular cluster. This issue is as yet undecided. Should it be confirmed, however, that open clusters instead of globular clusters are formed during collisions of galaxies, then it would be difficult to maintain that collisions are the source of elliptical galaxies.

Intense star formation in a nearby galaxy is another phenomenon where Hubble provided precise information for the first time. At a distance of 8 million light years, NGC 253 is the closest "starburst galaxy," where stars are forming at an extraordinarily high rate. Though this has been known for some time, only Hubble's high resolution enables us to speculate about the processes of star formation, by showing us the complex interplay of dense gas and dust with bright, extremely compact, young star clusters. Processes that are normally distributed over many places in a galaxy and over billions of years, occur here simultaneously and in a region only 100 light years across.

Top: NGC 7252 is the result of a collision of galaxies, left as seen from the ground, and right in a Hubble image (Source: left: F. Schweizer, right: B. Whitmore and NASA).

Right: This false-color image clearly shows newly formed conglomerations of stars (Source: B. Whitmore and NASA).

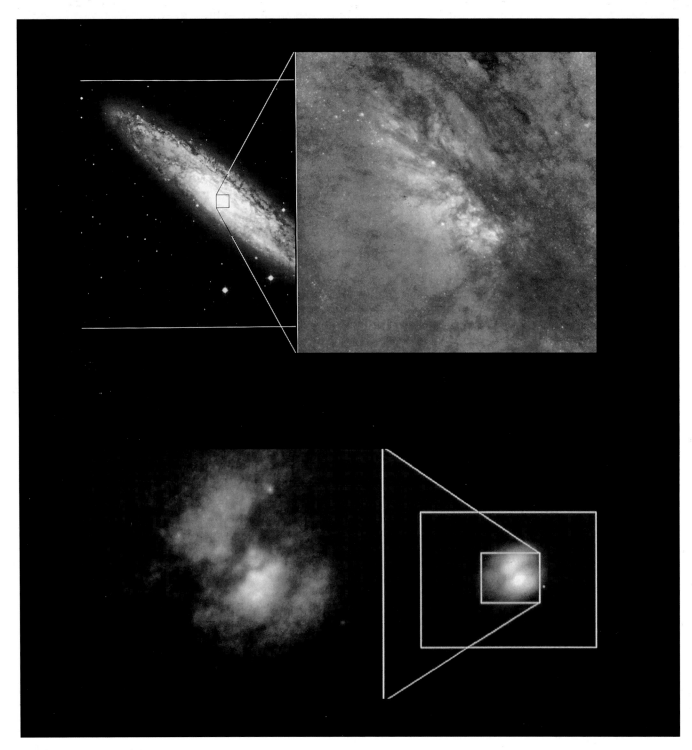

Top:
In this nearby galaxy (NGC 253) an unusual number of new stars are being formed – this is a "starburst" galaxy. Left: a ground-based image of the entire galaxy (Source: Carnegie Institutions); right: Hubble image of detail (Source: Jay Gallagher, Alan Watson, and NASA).

Bottom:
Arp 220, another starburst galaxy. Six bright areas of intense star formation result from the merger of two galaxies. These clusters are much brighter and ten times as large as other known star clusters. A dense dust lane appears to divide Arp 220 into two parts, but in reality it is a single galaxy; left: Hubble image, right: ground-based image (Source: E. Shaya and NASA).

Many gravitational lenses are at work in this rich cluster of galaxies, Abell 2218; they cause distorted images of galaxies at five to ten times greater distances. These images can be seen as arcs between the galaxies. Without this lensing effect, which also brightens the distorted images of the more distant galaxies, they would be invisible even for Hubble (Source: W. Couch et al. and NASA).

Gravitational Lenses in the Universe

Every child knows how a lens works. Rays emanating from an object can be focused by a convex lens and produce an image of the object on the other side of the lens. This is the basic principle of a magnifying glass (or a telescope). Lenses of a gigantic kind exist in space. They are not made of glass but of space itself. This is how they work. According to Einstein's general theory of relativity, the presence of matter curves the structure of space. A ray of light propagates along a straight line in space free of matter, but when passing a celestial object it travels along a slight curve. The bending of light in the gravitational field of the sun had been one of the predictions of the general theory of relativity; it was observed for the first time during a solar eclipse in 1919, providing dramatic confirmation of Einstein's theory. If two stars are positioned exactly behind one another in our line of sight, the light of the more distant star should appear as a ring around the point-like image of the closer star. In addition, the light is amplified, increasing the total intensity of the pair of stars. Since this effect is comparable to a lens, it is called a gravitational lens. Single stars can act as lenses, as can whole galaxies and galaxy clusters. The first gravitational lens was demonstrated in 1979–a double image of a quasar caused by a galaxy. One of the most impressive images of a gravitational lens is that of another quasar, called the "Einstein Cross" because of its shape and in honor of Einstein. Since then an intense search for further events of gravitational lensing has been under way–for quasar images, in the Magellanic Clouds, and in the direction of the center of our own galaxy. The reason for the latter two projects is the search for lensing from very faint objects, so-called brown dwarfs, which are midway in size between planets and stars. Lensing by brown dwarfs would be expected to lead to characteristic short-term brightness fluctuations in background stars as they

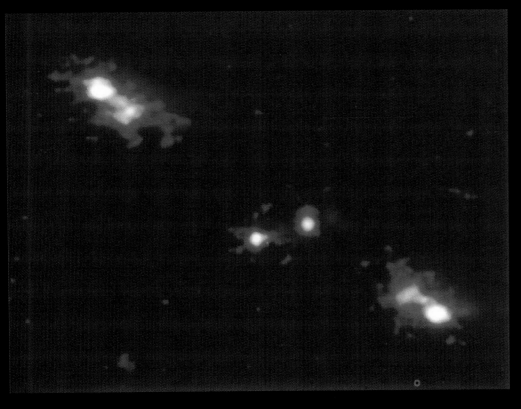

Top:
The gravitational lens as telephoto lens: The galaxy cluster AC 114 images a distant galaxy as two symmetrical objects (the L-shaped spots top left and bottom right). The two compact objects between the lensed images have nothing to do with this phenomenon (Source: R. Ellis and NASA).

Bottom:
The Einstein Cross is a particularly exotic example of a gravitational lens. A galaxy in the center causes four separate images of a distant quasar (Source: NASA).

move in the dense regions of the Magellanic Clouds or the galactic center.

Lensing is much more common with extended galaxy clusters than with stars. Such clusters curve space and produce ghostly images of galaxies located behind them and much farther away.

The Hubble image of the rich galaxy cluster Abell 2218 is a spectacular example of the effects of a gravitational lens, allowing us to see directly the effects of warping or curving of space. The pattern of arcs extending over the entire image like a spider web is due to the lensed images of a population of galaxies at a distance of five to ten times that of the cluster. Their light was emitted when the universe was only a quarter of its present age. Therefore, these images provide the opportunity to probe the early development of galaxies.

The picture also shows multiple images, which occur when the space curvature is strong enough. From the structure of these images astronomers can determine the distribution of matter in galaxy clusters – an interesting task, as such clusters should contain a large quantity of dark matter.

Finally, Hubble helped unmask a seemingly extraordinary galaxy in the distant universe whose physical parameters looked extreme in every way. But Hubble's images (as well as excellent ground-based images) showed that we are actually viewing this galaxy through a gravitational *microscope*. A foreground galaxy magnifies the intensity of the distant galaxy by a factor of 30 to 100. The arc-like shape of the distant galaxy's image fit gravitational lensing models per-fectly. A significant feat in its own right, this discovery also reminds us that not everything we see in the sky is true: gravitational lenses may distort our view.

Are Quasar Theories Out of Date?

In Chapter 1 we emphasized Hubble's ability to look into the deepest parts of space, and therefore into the distant past. These regions contain the most puzzling objects in the universe – objects whose very name reflects uncertainty: quasistellar objects, or quasars.

Since their discovery in 1963, quasars have been very poorly understood. If we start from the fact that the redshift in their spectra is caused by the expansion of the universe, then their extreme redshifts indicate great distance and intense luminosity. Since they are hardly found at low redshifts (which means close to the present and in the vicinity of our galaxy), they must have become extinct many millions of years ago.

One theory to explain their high energy production goes like this: Quasars are the cores of galaxies containing black holes with masses of several million suns. During the early phases of the universe, the center of such a galaxy contains plenty of stars, gas, and dust, which are captured by the black hole. This process produces a bright disk of accreting material around the black hole – the quasar. In later times, the supply of material diminishes, the black hole "starves," and the quasar fades.

Perhaps our own galaxy contains such a massive black hole at its center. If so, hardly any of its radiation would reach us. The nature of a possible mysterious

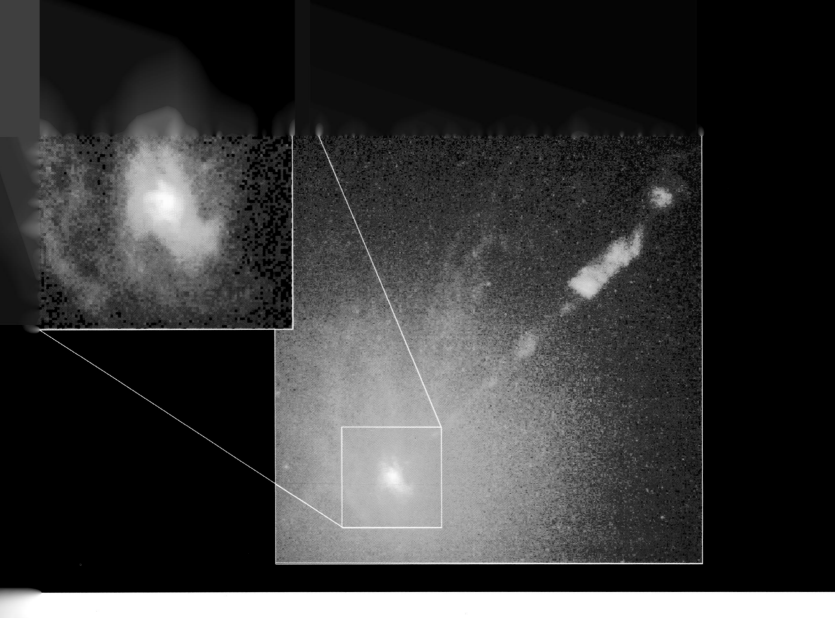

object at the exact center of the Milky Way is still being hotly debated. For some, the spectrum of the radiation escaping from this region is clear proof of a black hole. Others see it as indicating only a star with an unusual accretion disk. The so-called Seyfert galaxies are transition objects, normal spiral galaxies with unusually bright cores but less extreme than quasars. Ground-based astronomers have tried in recent years to take pictures of the faint, small galaxies purportedly surrounding the bright quasars, but the turbulence of the atmosphere and the high brightness contrast between quasar and galaxy have made those observations very difficult and often unconvincing. Only the Hubble has provided good images of quasars. Do they confirm the conjecture that quasars are central objects of normal galaxies? Do they finally present proof for the existence of black holes?

Before we turn to Hubble's observations of quasars, we first have to discuss images of active galaxies. The giant elliptical galaxy M87 is the brightest object in the Virgo cluster, at a distance of about 50 million light years. A jet emanating from the core region, made up of fast electrons and positrons, has been known to exist for some time; it can be seen

The active galaxy M87, complete with a jet of electrons and positrons close to the speed of light, and with a central gas disk. Does this galaxy harbor a supermassive black hole? (Source: H. Ford et al. and NASA)

in great detail in the right picture above. But the nature of the galaxy's bright core had not been certain. It was assumed that there would be a black hole, a central "engine" responsible for powering the bright core and the ejection of the jet. Using Hubble images, scientists were able to examine the central area more closely. A spiral cloud of gas can be seen around the core (see insert on the left). This gas produces emission lines, so that its velocity can be determined from the Doppler shift of those lines. The gas in the disk circles the core with a speed of 550 kilometers per second; from the size of the disk and the orbital period, astronomers can determine the mass of the central object. A core made up of stars would hardly be able to exert such gravitational forces, and it would look different as well. The best explanation for what we observe is that at the center of M87 there is a giant black hole with several billion times the mass of the sun.

The giant elliptical galaxy NGC 4261 is one of the brightest in the Virgo cluster. In visible light, the galaxy appears as a nebulous disk, consisting of several hundred billion stars. The picture shows the core of NGC 4261. A huge disk of gas and dust "feeds" a (presumed) central black hole. This disk extends over 300 light years and is inclined at an angle of about 60 degrees, so that we can see the bright nucleus at the center of the disk. This is the immediate vicinity of the black hole. Perpendicular to the disk, hot gas is being ejected from the vicinity of the black hole, giving rise to the radio jets. In late 1995 it was announced that the velocity field inside the dust and gas disk of NGC 4621 had been measured by detailed Hubble spectroscopy

in two dimensions. The observations were consistent only with orbital motion around a central mass of roughly a billion times the mass of the sun – presenting the second strong case for a supermassive black hole in a galaxy, after M 87. Critics, however, maintain that other astronomical objects (admittedly only slightly less bizarre) could also account for such a large mass in such a small volume – for instance, a supermassive disk. Meanwhile, dramatic dark disks in elliptical galaxies have turned out to be a frequent phenomenon.

The barred spiral Seyfert galaxy NGC 5728, at a distance of 125 million light years in the constellation Libra, shows a remarkable double cone of light in its core. Here, the opaque disk appears at an angle that makes the central object invisible. We can only see matter above and under the disk, which is being stimulated to emit light by the energetic radiation emanating from the central object.

The galaxy NGC 1068 is the prototype of a class of objects known as Seyfert-2 galaxies. Such galaxies have very bright cores, radiating with a luminosity of a billion suns. The brightness of the core varies on time scales of a few days, indicating that the emitted energy is generated in a region with a dimension light can traverse in a few days – hardly larger than our own solar system. Only a black hole of 100 million solar masses, capturing matter from its vicinity, appears to be able to provide such energies. In the case of NGC 1068, earlier Hubble observations had shown a number of hot gas clouds being heated by the energetic radiation from the central source. The

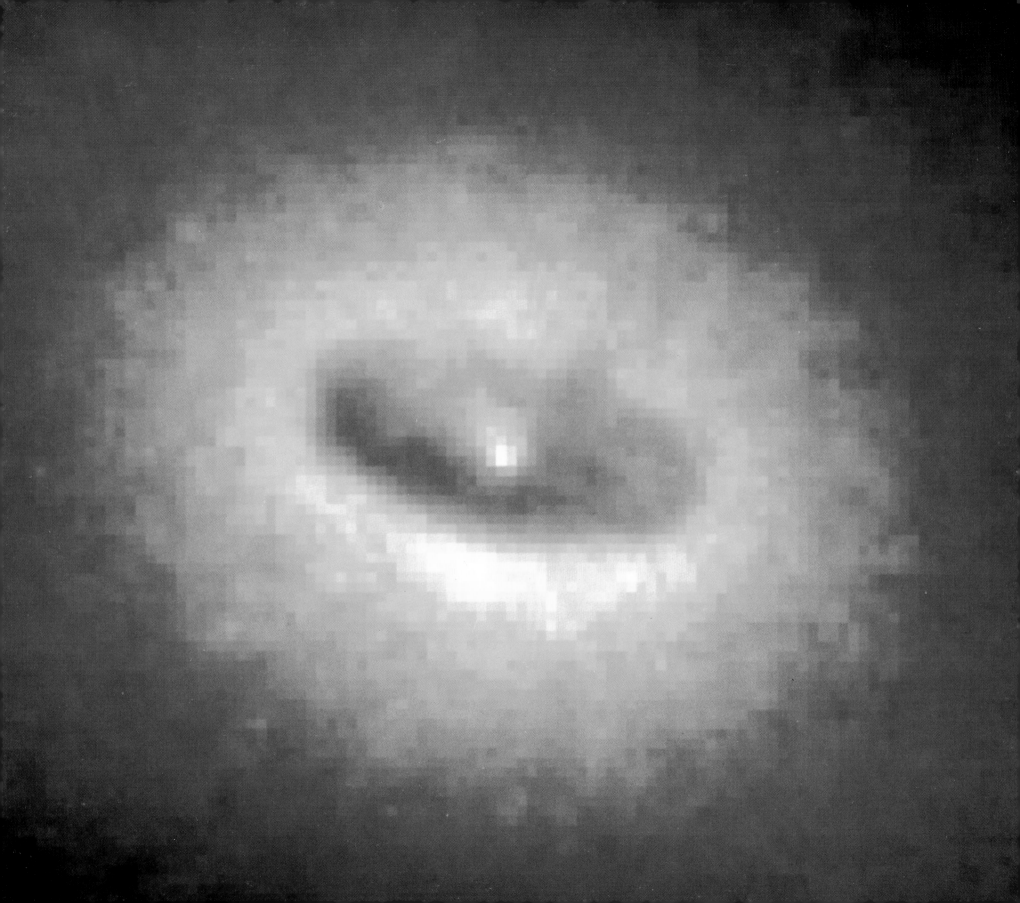

Left page:
Did Hubble find a black hole? Many astronomers believe such an object is located in the center of the galaxy NGC 4261, hidden behind a dark gas and dust disk (Source: W. Jaffe, H. Ford, and NASA).

The Seyfert galaxy NGC 5728 and its interesting center: A dark ring around a "central engine" appears to exist here as well. Left: the entire galaxy; right: the core region and light cone (Sources: A. Sandage / A. Wilson et al. and NASA).

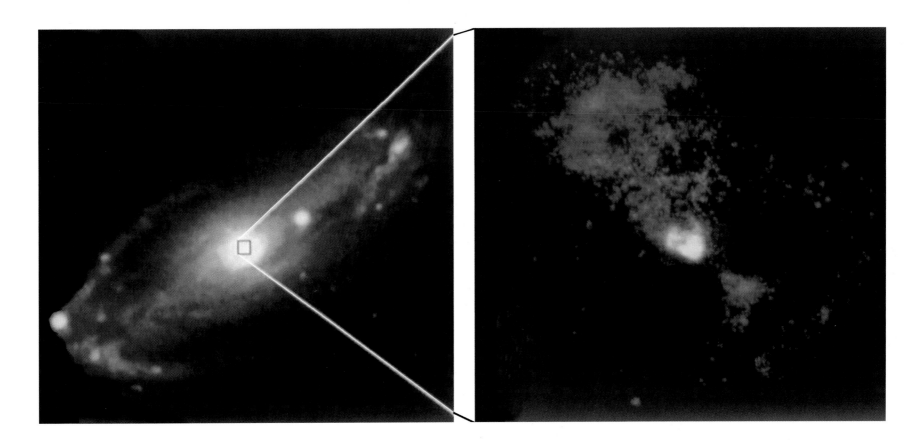

new observations with the Faint-Object Camera and the COSTAR corrector show an extended emission region, made to glow by the radiation from the central object, and containing hitherto unknown and unexpected filamentary structures.

Because the most luminous quasars are much more distant, it is more difficult to examine their surroundings. However, this is a problem Hubble can surmount. The results were surprising. The picture shows the quasar 1229+204, situated in the core of a barred spiral. This galaxy is undergoing a collision with a dwarf galaxy, leading to increased star formation and to an improved supply of "fuel" for the quasar. An extended blue region,

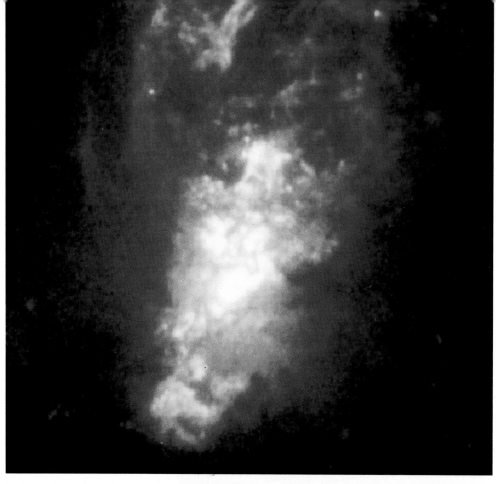

clearly visible on one side, probably consists of massive young star clusters formed as a result of the collision. Shell-like structures along the bar might be due to tidal effects between the spiral and the companion galaxy.

Top:
The core region of the Seyfert galaxy NGC 1068 (Source: D. Macchetto et al. and ESA).

Up to this point, scientists had assumed that quasars would always be found in the center of a bright "host" galaxy, devouring stars and gas in their vicinity. It thus came as a shock to the astronomical community when, early in 1995, leading Hubble observers announced that they had failed to locate *any* around a large percentage of observed quasars. Was the standard model sinking? Certainly not: this mystery didn't last a year. Other researchers, using different methods of image processing, had no trouble finding host galaxies for each quasar studied, though the types and sizes of these galaxies didn't always match expectations. By the end of 1995 it was generally accepted that quasars do indeed reside in galaxies – but not, as had always been thought, in particularly bright ones. Either the host galaxies are relatively normal despite the presence of a bright quasar, or they are tidally disturbed or faint and highly disrupted structures. How we account for this is another question, though one rudimentary model proposes that quasars result from collisions between spirals and other types of galaxies. Not every colli-

Bottom:
Here, the standard model appears to be correct: A quasar (QSO 1229+204) is located in a faint host galaxy (Source: J. Hutchings and NASA).

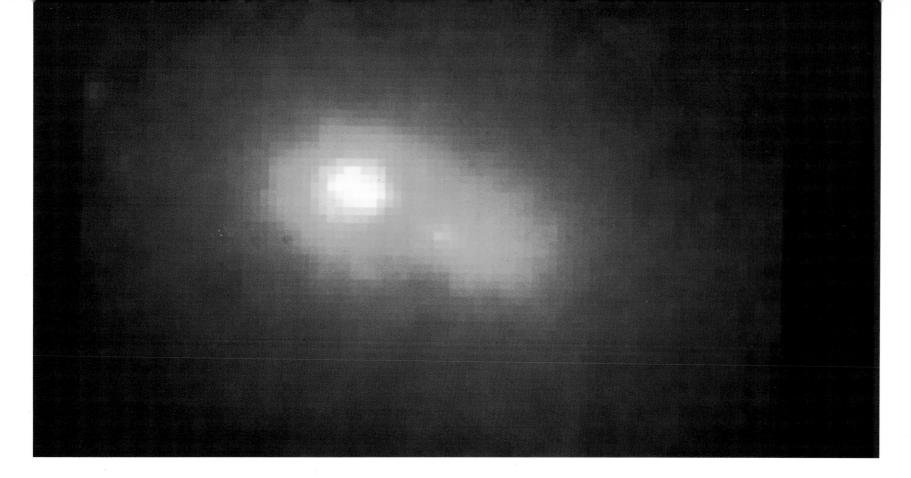

The cause for many speculations: the mysterious core of M31 (Source: T. Lauer and NASA).

sion, however, produces a quasar. The situation in this field is characteristic of Hubble science in general: five years after launch, the telescope delivers large amounts of astonishing data that will rewrite the textbooks. But we don't know how it will rewrite them.

Mysterious Galaxies

Surprise was also the rule with the mysterious double nucleus of the Andromeda galaxy (M31). This is the closest large spiral galaxy, a true mirror image of our own Milky Way, and "normal" in every way. At least that was the opinion until 1993, when Hubble observed its core region with high resolution. Hubble found not one but two bright spots, only 5 light years distant from each other. The center of mass, around which all stars in the galaxy revolve, is not even the brighter spot, but the much fainter of the two! Although there are several ideas as to why this might

be, a satisfactory answer has not yet been found. There are speculations that M31 could have recently swallowed another galaxy, or that a thick dust layer produces just the illusion of a double nucleus. Or maybe an asymmetrical disk around the true nucleus of M31 produces the second spot.

Why is the center of the galaxy M51 marked with a cross? Whenever Hubble observes the central regions of galaxies, surprises seem to occur. The close spiral galaxy M51, for instance, revealed a mysterious cross at its exact center. The darker bar is probably a dust ring around the central engine seen edge-on, hiding the processes in the innermost few light years. Radiation can escape only toward its poles. But the second, tilted bar evades all explanations: Another disk? Why don't they destroy each other, or merge after a short time?

One thing is clear: Hubble has shown that the cores of galaxies are much stranger than astrophysicists had thought.

Strange bars dissect the spiral galaxy M51 (Source: NASA).

Large and Small Lights: The World of Stars

Watching a magician, we are again and again amazed and delighted by the rabbits, doves, and flowers appearing apparently naturally from his hat or coat. But the things nature can conjure up surpass all imagination. From the simplest of ingredients, the chemical elements hydrogen and helium, it is able to form the most diverse objects – nebulae, stars, and systems of stars. And not all stars are similar. Particularly if there is time for evolution, they appear in great variety. There are stars smaller than Earth, and there are stars several hundred times larger than the sun. There are stars almost as old as the universe, and others younger than humanity. They can be single or double stars, or live in large families called star clusters. They can appear as red giants, white dwarfs, neutron stars, or black holes. They can pulsate, explode suddenly, or flicker like a defective fluorescent tube. The Milky Way alone is home to 200 billion stars – more than there are humans on Earth. It is obvious that the exploration of stars is one of Hubble's most interesting tasks.

The Most Massive Stars

For us earthlings one star is of prime importance: the sun. It is the closest star to Earth, it is circled by Earth and all other planets, and without its light, which it has been warming Earth for 5 billion years, life could not have developed on our planet. But for astronomers the sun is a totally ordinary star in a not very interesting part of the Milky Way.

In some ways stars are like humans: there are extremely large and heavy ones, and there are small and light ones. For science, both extremes are of interest – the dwarfs and giants in the sky. The sun, with its mass of 2×10^{30} kg (a 2 followed by 30 zeros, corresponding to 300,000 Earth masses) and its diameter of 1.4 million kilometers (corresponding to 109 Earth diameters), is only an average star, in spite of these imposing numbers. The most massive stars contain about 100 solar masses of material, and the least massive 0.1 solar mass. Massive stars use up their nuclear fuel much faster than the sun, so they radiate their light into space for only a few million years before running out of hydrogen. Nuclear reactions turn the elements in their interior into successively heavier elements, and when the core is converted to iron they collapse and light up for one last time as a supernova. Short-lived, massive stars are rare inhabitants of space – they can be found only in regions of dense interstellar matter, in young star clusters partly shrouded by glowing nebulae, since they don't live long enough to travel far from their places of birth.

One object that had been puzzling astronomers for a long time is called R136, a bright spot of light in a star cluster of the Large Magellanic Cloud. The distance of this object is known very accurately, and therefore its luminosity, or energy released into space, can be calculated. The mass necessary to release this much energy turns out to be several hundred solar masses. Is this monster perhaps the most massive star?

Before the launch of the Space Telescope, there were already indications that R136 might not be a single star but a densely packed cluster of bright, not extremely massive stars. This was a good question for Hubble to answer. One of the first images indeed showed that R136 was a compact cluster of stars. The best images demonstrate that it is a collection of more than 3000 stars!

This compact star cluster is located in an area of star formation, a region of ionized hydrogen gas, known as 30 Doradus. All stars and gas clouds visible there are part of the Large Magellanic Cloud, the closest neighbor of the Milky Way. From these stars, light takes 160,000 years to reach us.

Although 30 Doradus may appear relatively small, we have to remember that this collection of newly developing, just formed, and almost burned-out stars is many hundred times more distant than the apparently brighter and much more famous region of ionized hydrogen called the Orion Nebula. While the Orion Nebula is illuminated by only a handful of stars, 30 Doradus contains hundreds of stars, which emanate sufficient amounts of energetic radiation to light up the nebula.

Ground-based telescopes are able to discern details in 30 Doradus of a size of about 200 light days (the distance which light travels in 200 days, or just over one half of a light year). Hubble sees details with a resolution of 25 light days, almost ten times sharper.

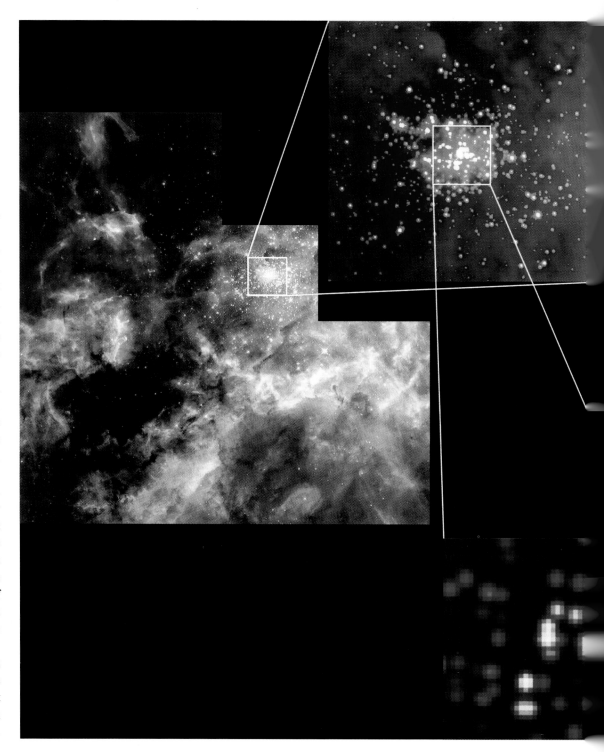

The 30 Doradus nebula, a spectacular region of star formation in the Large Magellanic Cloud, a neighboring galaxy to our own Milky Way. Left: A complete mosaic of the nebula. The enlargement shows the compact star cluster R136a, resolved by Hubble into more than 3000 stars; a further enlargement shows the brightest stars in the central region of the cluster (Source: NASA Goddard Space Flight Center).

For comparison, the diameter of the solar system is half a light day, whereas the closest star to the sun, Alpha Centauri, is at a distance of 4.3 light years.

Theoretical calculations show that the maximum mass for a star with stable energy output, fueling its radiation by nuclear fusion of light elements into heavier ones, can be up to about 100 solar masses – but theories always have to be tested by observations. As we know today, R136 is not the object to test these theories. Some Hubble observations of bright stars in other galaxies suggest something is lacking in our understanding of the upper mass limit of stars. Judging by the strong stellar winds coming from some extremely luminous stars, astronomers have derived sizes of up to 200 solar masses. Hubble's sharp vision can at least exclude the possibility that these are actually close binary systems, but critics have attacked the physical assumptions underlying these mass calculations. This is yet another area where the Space Telescope is challenging astrophysical theory.

The Least Massive Stars

What about the low end of the scale? Low-mass stars are also cool stars: Nuclear processes occur at a slow rate and heat the stellar surface to only a few thousand degrees. By comparison, the sun's surface is about 6,000 degrees. These cool, small stars are called M dwarfs. Because of their low luminosity, they cannot be detected outside the close vicinity of the sun.

One of the faintest stars is Gliese 752, a double star at a distance of 19 light years in the constellation Aquila. The brighter component, Gliese 752A, is a red dwarf star with a mass 30 percent that of the sun. Its fainter companion, Gliese 752B, has only about 9 percent of the sun's mass and a diameter smaller than the planet Jupiter. The theoretical lower limit for a true star, producing its own nuclear energy, is estimated to be 8 percent of a solar mass. Jupiter has a mass of only 0.1 percent of the sun's. Its interior cannot sustain the nuclear reactions typical of stars, so it remains cool and mostly radiates only the light received from the sun. Between the least massive stars and the planets like Jupiter could exist the brown dwarfs, collections of matter with mass insufficient to ignite the nuclear processes in their interior. How many brown dwarfs are there? We don't know, since their observation is extremely difficult. Why are they important? In spite of their diminutive size they could be important for the structure of the universe. If they existed in large numbers, they could be the explanation for the amount of matter necessary for the density of the universe. This "dark matter," diligently looked for by cosmologists, could possibly fill the universe in the form of such brown dwarfs or Jupiters that have remained undetected, although the likelihood of this seems small as they appear to be uncommon. Recently, however, Hubble appears to have found a genuine brown dwarf, a companion to the star Gliese 229, 19 light years distant in the constellation Lupus, with a mass about 20 to 50 times that of Jupiter.

Gliese 229B bridges the gap between the smallest stars, which must have at least 80 Jupiter masses to shine, and the largest planets, with which it shares important chemical characteristics such as methane in its atmosphere.

Low mass stars show brief, dramatic increases in brightness caused by the release of stored magnetic energy. These so-called stellar flares occur on all cool stars. They can be observed in detail on the sun, on stars of the solar neighborhood, and on faint stars in young star clusters. The fainter the star, the more spectacular such flares appear, since they stand out more.

One of these flares was observed in October 1994 on the star Glie se 752B. A one-hour exposure in the ultraviolet spectrum showed no significant signal during the first 55 minutes. But during the last 5 minutes a flare occurred, heating the outer layers of the star to a temperature of 150,000 kelvin. Flares on extremely low-mass stars are remarkable, since they indicate that phenomena occur that are similar to those on the sun. However, we don't know whether flares might occur on brown dwarfs, perhaps alerting to their existence.

Another low-mass star is Gliese 623, at a distance of 25 light years. Again, it is a double star, but the less massive component has never been observed with a ground-based telescope. The movement of the brighter component around the common center of gravity causes a minute periodic oscillation of the visible star, which gave astronomers an indication

that a low-mass companion might exist. The orbital period around the common center of gravity is 4 years, and the distance of the two stars from each other is twice the distance from Earth to the sun.

Using Hubble, the faint companion was observed directly for the first time, and thus it became possible to derive temperature, brightness, and mass, providing improved knowledge about very low mass stars. This object has only one-tenth the sun's mass and radiates only 1/60,000 the amount of light. If it were to take the place of the sun, our days would hardly be brighter than a night at full moon now.

Families and Clans of Stars: Open and Globular Clusters

Only rarely do stars exist as only children. We assume that the majority, including our sun, are formed in groups, in open and globular clusters. Open clusters dissolve over billions of years, losing more and more stars to the extended, flattened disk of the Milky Way.

But if a star cluster is more compact and more massive, its life expectancy is longer. The oldest clusters may contain stars with an age of 10 billion years or more. These objects began their existence when the universe was still relatively young, and when the interstellar clouds from which these stars are formed had not been enriched by dust from previous generations of stars. These massive globular clusters of our galaxy are made up of old, metal-poor stars (astronomers have the peculiar habit of referring to all the elements

Gliese 623b, one of the faintest stars, with only one tenth of the solar mass and 1/60,000 of the sun's luminosity. In this image it is marked by an arrow, close to the more massive star which it orbits at a distance of 200 million kilometers (Source: C. Barbieri et al. and NASA/ESA).

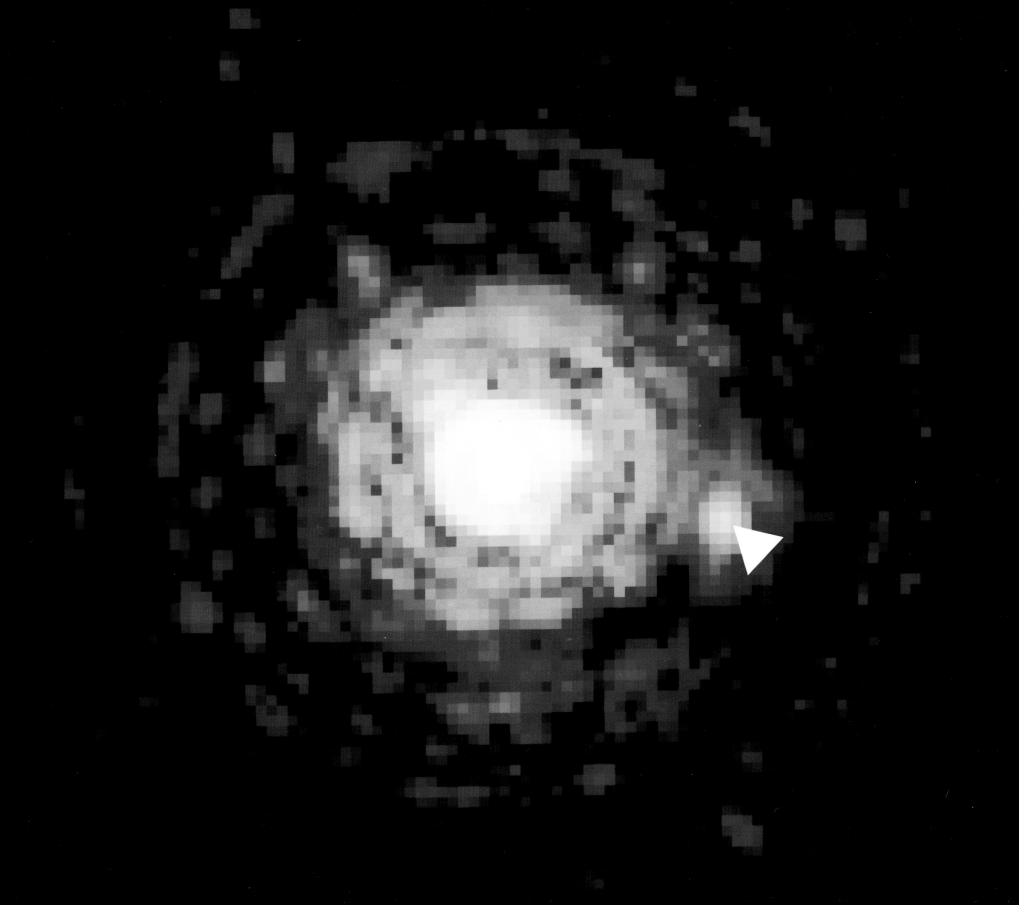

Left page:
Hot blue stars in the core of the globular cluster M15 (Source: G. de Marchi/F. Paresce and NASA/ESA).

Two large star clusters in the Large Magellanic Cloud, resolved into individual stars by Hubble. The dominant yellowish cluster is NGC 1850 (Source: R. Gilmozzi et al. and NASA).

besides hydrogen and helium as "metals.") From these stars' intensities and colors, their age, and the age of the cluster, can be determined from comparisons with computer models. One finds that our galaxy contains both old and young open clusters but, within a narrow range, only old globular clusters.

It remains unclear whether our entire galaxy was formed from such globular cluster "building blocks," with the currently existing clusters representing the remaining materials, or if the old parts of the galaxy – the halo and the extended central part surrounding the core – and the globular clusters all formed at the same time. Both the old parts of the Galaxy and the globular clusters have an age of about 15 billion years. Are there young globular clusters as well? Yes, but not in our galaxy. It is a worthwhile task, therefore,

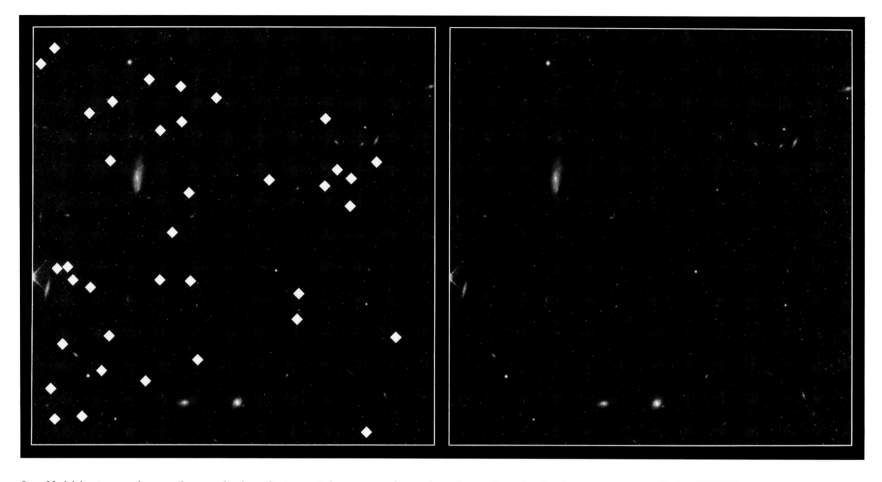

for Hubble to explore other galaxies that contain comparatively youthful clusters, like the Magellanic Clouds.

The Hubble telescope was able to detect a previously unknown type of extremely blue stars in the globular cluster M15. In the central parts of this stellar aggregate the density is so high that stars strip away their outer thin layers from each other, a process that has been called "stellar cannibalism." The 15 blue, hot stars observed by Hubble are reduced almost to their stellar cores, with the outer parts gone. M15 is distinguished by having the highest central star density known – even Hubble can't resolve its innermost core. Star counts on Hubble images of M15's heart suggest that this cluster may have

experienced a long-theorized phenomenon called "core collapse," or "gravothermal catastrophe." This means that on a cosmically very short time scale of just millions of years, many stars lost most of their orbital energy to outer parts of the cluster and now huddle close to the center, although other mechanisms prevent actual stellar collisions. In theory, such core collapses should occur in about one in five globular clusters, but the observational evidence is nowhere stronger that in M15.

In spite of a distance of 166,000 light years, it is no problem any longer for the repaired Hubble to resolve two star clusters in the Large Magellanic Cloud into their individual stars and to show that there exist not one but two clusters positioned behind

An arbitrary piece of sky, 1.5 arcminutes on the side – and surprisingly poor in faint stars! The image on the left indicates the expected density by artificial symbols. Their absence means that faint red stars cannot account for a significant percentage of the dark matter in our galaxy (Source: J. Bahcall and NASA).

one another along the line of sight. About 60 percent of the stars belong to the dominant yellowish star cluster NGC 1850, about 50 million years old, but the dispersed white stars are part of another cluster only 4 million years old, and about 200 light years farther away. Perhaps supernova explosions in the old cluster contributed a gas cloud that helped form the young cluster.

The color image on page 99 is a composite from three exposures in different wavelength ranges: Yellow stars correspond to "normal" stars (so-called main sequence stars) with surface temperatures of about 6,000 kelvin, similar to our sun; red stars are cool giants and supergiants with temperatures of 3,500 kelvin; and white stars are hot young objects with temperatures of 25,000 kelvin and more.

Finally, thousands of stars have to be mentioned, which have not been found by Hubble despite an intensive search. According to current theories of stellar formation, our galaxy should be full of red dwarfs, very small, cool stars barely able to generate energy from nuclear reactions. This population of extremely faint stars was a potential candidate for the explanation of one of the puzzles of modern astrophysics: Why does the outer part of our galaxy (and of many other galaxies as well) rotate faster than would be expected from the Keplerian laws of classical celestial mechanics? If we disregard the somewhat exotic interpretation that these laws themselves are incomplete, we are left with the hypothesis that there is additional mass in the outer parts of the galaxy, mass whose gravitational pull causes the faster rotations. The postulated red dwarfs would have been ideal candidates. Hubble should have seen them after its optics were improved – but they weren't there! The image on the right shows a typical field in the sky, as actually observed by the Space Telescope, while the simulated image on the left indicates the number of red dwarfs that should have been visible. Obviously, nature refuses to form stars with less than one fifth of a solar mass. Stellar theorists have to reconsider their theories, and the hunters of dark matter have to look for new candidates!

Gas and Dust: Of Life and Death of Stars

In the previous chapters we have dealt with galaxies and with stars inhabiting them. But what lies between the individual stars? Here, in seemingly empty space, we find the material from which new suns can coalesce. This material may be thinly distributed or collected in giant clouds, or, most remarkably, in radiating nebulae of hydrogen and other gases called HII regions. A single gas cloud can only emit radiation of low energy, in the radio window. Normally, it is not hot enough to produce more energetic, shortwave radiation. However, a cloud in which young stars have formed through condensation provides a fascinating view and is a frequent target for astronomical imaging. The shortwave, energetic radiation of these young stars excites the gas of the cloud, so that it glows. But how does a cloud "glow"? The hydrogen atoms of the cloud are ionized – stripped of their electrons by the stellar radiation; as the electrons fall back to the atomic nuclei, they emit light at well-defined and characteristic wavelengths. Astronomers refer to ionized hydrogen as HII, hence the name HII region for a cloud of (predominantly) ionized hydrogen. Such a region usually appears in color images as a reddish, frayed nebula; as the light is dispersed into the colors of the rainbow by a spectrograph, the bright line spectrum of a hot gas can be seen, similar to that of a fluorescent tube.

How do hydrogen clouds form? One has to imagine a dynamic process for their development: The beginning is a very extended "molecular cloud," cool enough for atoms to form chemical compounds. Parts of this cloud collapse under their own gravity to form hot, massive stars. These heat the surrounding cool gas, which has already begun to decay into denser structures, and ionize it. The atoms lose their electrons, and an HII region is formed. Aside from gas, it contains dust particles made from silicates or carbon (graphite). This dust shields the dense molecular cloud against the potentially damaging ultraviolet radiation from the hot stars, and enables complex chemical processes in the cool gas. The dust causes the cloud to appear dark and mysterious to us.

The Orion Nebula: Nursery of Stars and Planets?

The HII region closest to us is the Orion nebula, already known in the sixteenth century, and visible with binoculars or even the naked eye. Since the end of the nineteenth century we have known that it is a giant gas cloud. A certain type of variable star, the so-called flare stars, as well as evidence for infrared sources prove that the Orion nebula has to be a stellar nursery.

The early stages of star formation take place deep in the molecular clouds, which are opaque to visible light; they can be observed only in the infrared and radio regions of the spectrum. Despite this, astronomers have a fairly good idea about the birth of new stars. First, part of a molecular cloud becomes unstable and starts to contract under its

Gas nebulae are colorful subjects, and Hubble shows them in previously unknown detail. Many of the compact bright spots are protoplanetary disks (Source: C. O'Dell and NASA).

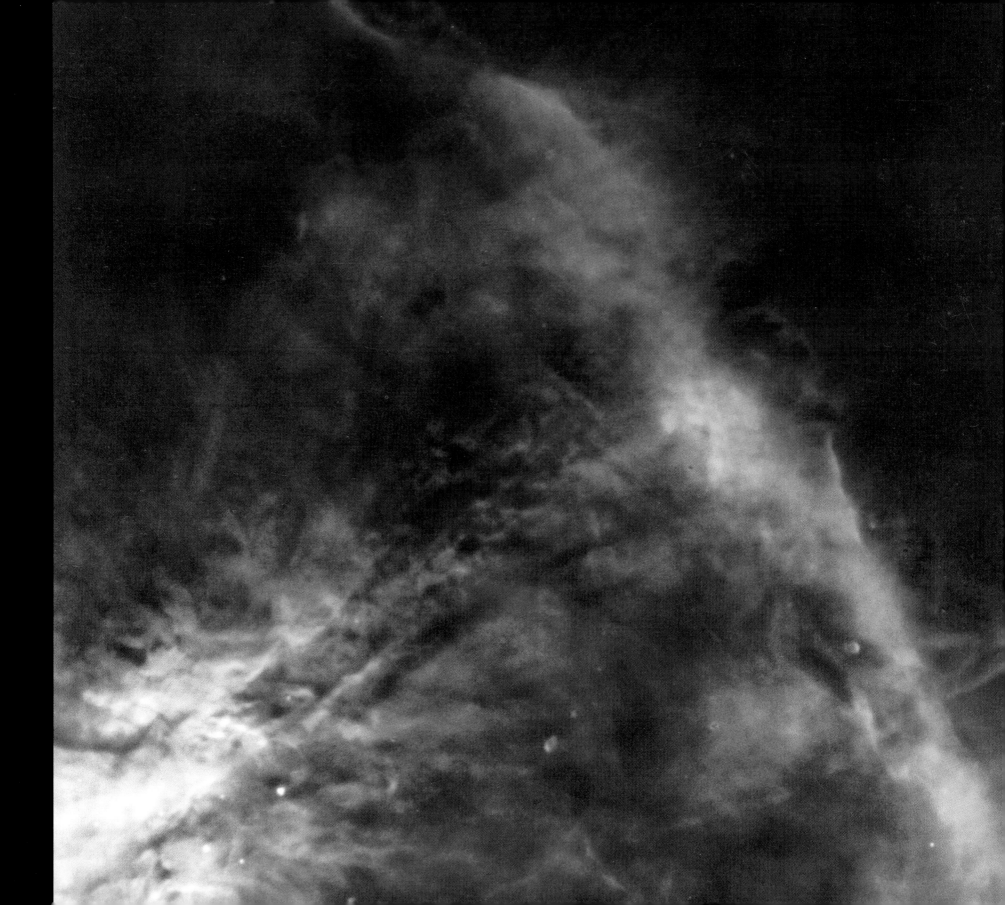

own gravitation. The core of the cloud, which will eventually become the new star, grows denser. At the same time, a disk of gas and dust forms around the central condensation; this disk will surround the equatorial region of the new star for several hundred thousand years, and may even form planets. We think this is how our own solar system and Earth itself could have formed. Evidence for rudimentary dust disks has been found even around "grown-up" stars. Stars in their infancy can be completely obscured by them.

The Hubble images of the Orion nebula give an impression of the variety of different processes taking place in its interior. The nebula itself is made to glow by young hot stars, visible in the upper part of the image on page 106. The brightest areas are protrusions from the surface of the nebula. The bright band on the left is caused by perspective: The terrestrial observer is looking along a long wall – a very long wall indeed, as the diagonal of the image corresponds to about 1.6 light years.

The image on page 107 shows a very small detail of the Orion nebula, including five very young stars, four of which are surrounded by gas and dust disks. These "protoplanetary disks," or "proplyds," may condense into planets during later stages of stellar evolution. The disks in the vicinity of the hot stars of the "parental" cluster appear to be bright, as they scatter their light; the most distant object, however, is insufficiently illuminated, so that it appears dark against the background of the Orion nebula. The images show an area of only 0.14 light years on a

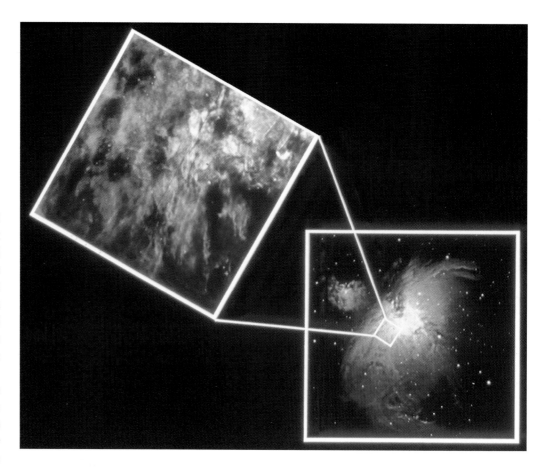

side. Only Hubble can provide so much detail about the proplyds at this time.

The extreme enlargements on the bottom of page 107 show three extremely young stars in the Orion nebula, at an age of only a few hundred thousand years, and still surrounded by materials left over from the process of star formation. The cool, reddish star on the right has only 0.2 solar mass. The proplyds appears as thick disks, with the actual stars in their center. Each image shows a field only 12 light days across.

The vast Orion nebula has become a hunting ground for phenomena relating to the birth of

The comparison shows the new dimension in Hubble's images; ground-based (right) and Hubble (left) images of the Orion nebula (Source: D. Malin and NASA).

stars. Two great surveys with the Hubble Space telescope have identified so many objects in various evolutionary stages that some statistical analysis is now possible. At least one out of every two stars in the nebula has some kind of circumstellar disk, most of them bright because they are ionized by ultraviolet radiation from neighboring hot young stars. In seven cases, however, the disks are black. The sheer volume of circumstellar matter surrounding the young stars in this nebula makes astronomers think large numbers of planets will someday form here.

Jets and Star Formation

Highly focused jets of matter are often seen emanating from the gas and dust disks around young stars. These jets are almost invisible in empty space, but as they slam into neighboring gas clouds they become very bright. At such a location, a luminous and strangely shaped nebula develops, called a Herbig-Haro object after the discoverers of this phenomenon. About 300 such Herbig-Haro objects have been discovered from the ground, but the limited resolution of ground-based telescopes posed restrictions on detailed understanding of the phenomenon. Basic questions about the physics of such stellar jets can be addressed only with the space telescope (pages 108–109). Does the apparent similarity to jets emanating from galaxies also mean that they are as fast (some of these galactic jets appear to approach the speed of light), or do they have velocities of "only" a few hundred kilometers per second? How close to the star do these jets form, and how are they held together, or collimated? And how uniform is the rate of ejection of matter?

The answers provided by the Hubble do not provide a uniform scenario. According to Hubble's data, jets form within the disks surrounding young stars early in their development and are focused immediately. Rather than the disk itself acting as a kind of nozzle, magnetic fields may be responsible for the strong collimation. The jets are not ejected uniformly, but they appear to be clumpy. This means that material from the disk falls toward the young star and is fed into the jet in an irregular pattern. There may be intervals ranging from a few months to several dozen years between infalls of matter. Furthermore, the direction of the jet does not appear to be fixed. The explanation for this behavior may be found in a potential companion star, around which the young star orbits. These and other interesting details exhibited by young stars have to be incorporated into new and improved theories.

Planetary Nebulae: A Colorful Finale

Gas clouds in space are not only associated with stars' formation, but also with their death.

Stars are formed from gas, and in their death struggle they relinquish a large part of their mass again into space, now enriched by heavy elements produced by nuclear fusion inside the star over its

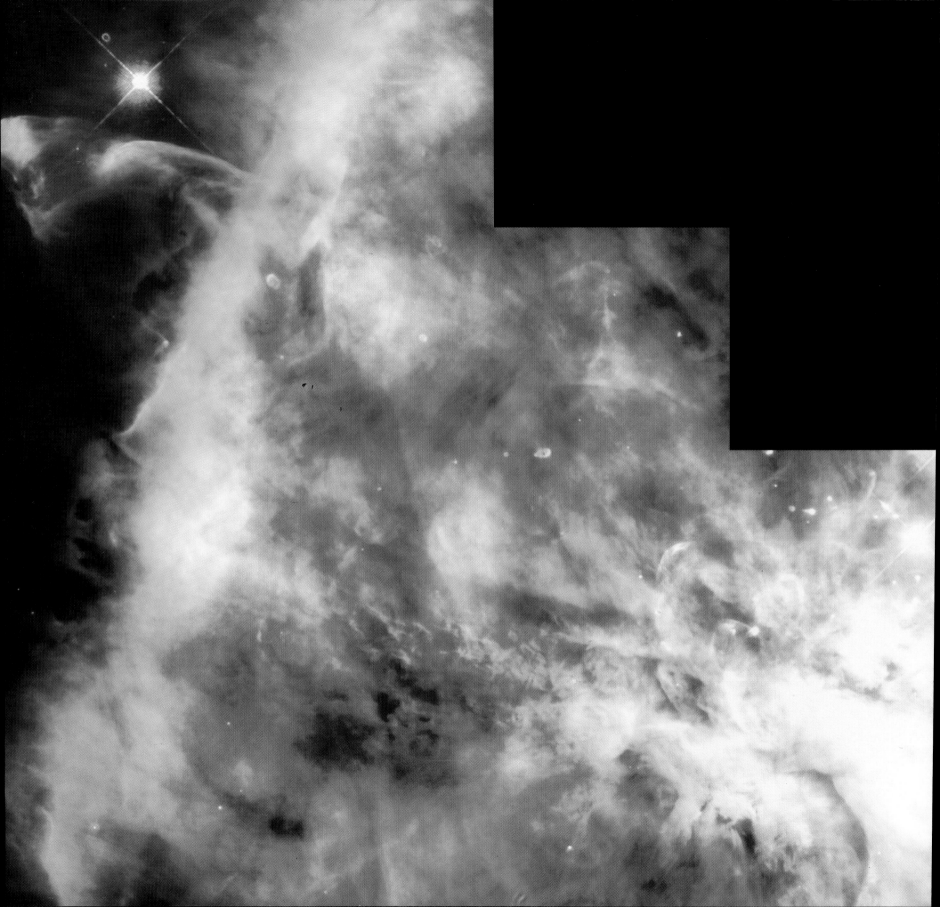

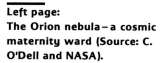

Left page:
The Orion nebula – a cosmic
maternity ward (Source: C.
O'Dell and NASA).

Five young stars in the Orion
nebula. Four of them are
surrounded by gas and dust
disks – matter from which
planets can form. The bottom
row shows such stellar babies
in detail (Source: C. O'Dell and
NASA).

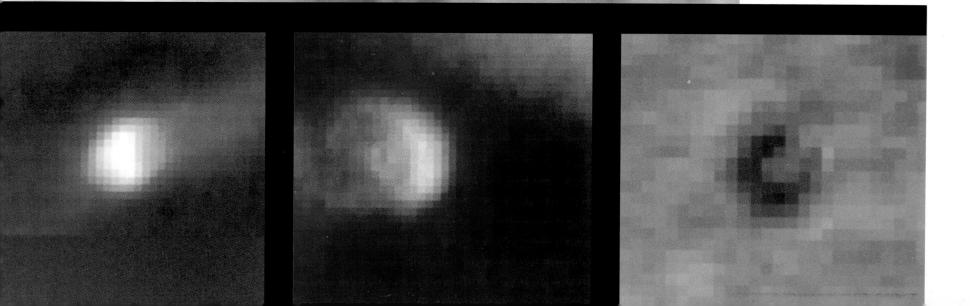

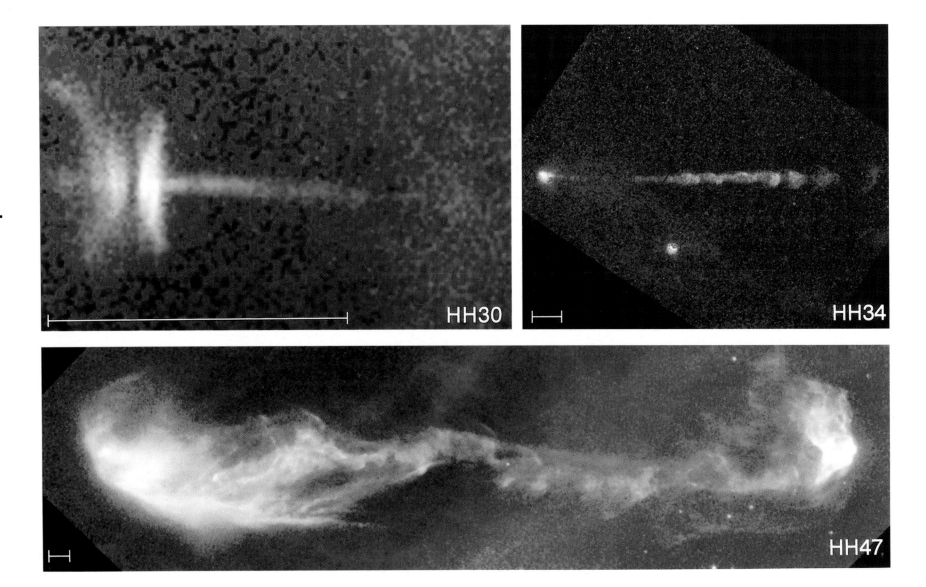

HH30

HH34

HH47

Spectacular images of focused beams of matter, ejected from gas and dust disks around young stars:
top left shows Herbig-Haro object 30, where the disk hiding the ejecting star appears edge-on. The jet which emanates from the disk is already well focused.

On the right: HH-34 (Sources: C. Burrows / J. Hester and NASA).
In the bottom image we see HH-object 47: This time, the responsible young star hides in a dust cloud on the left edge of the image. The jet follows a complex pattern. The jet is about 150 billion kilometers

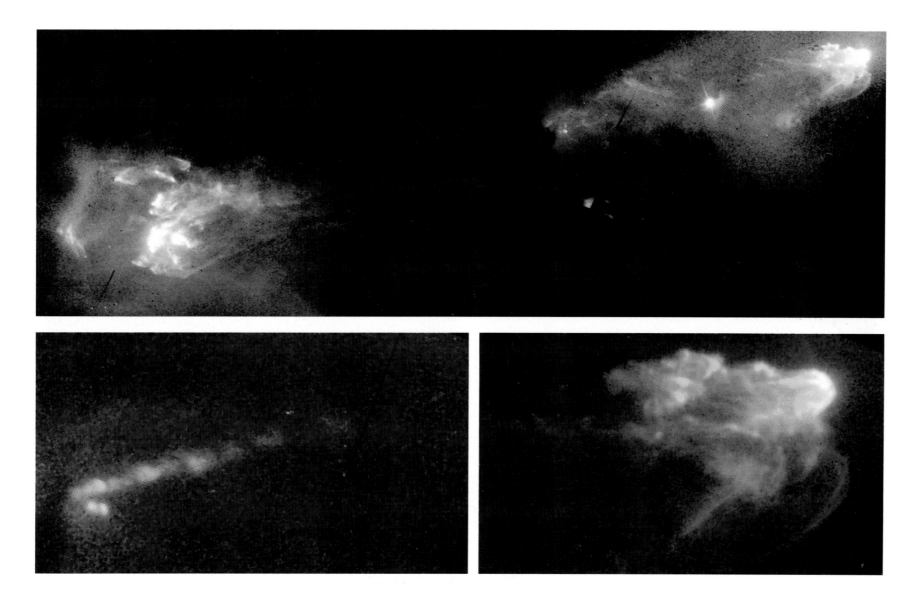

long, 1000 times the Earth-Sun distance (Source: J. Morse and NASA).
In the top image, the star is located in the center, but hidden by dust. The almost symmetrical bubbles indicate the areas where the jets penetrate the interstellar gas; they are called HH 1 and HH 2. The detail on the bottom left shows part of the gas jet and indicates that it does not stream uniformly, but intermittently. The enlargement on the bottom right shows the classical structure of a bow shock at the place of collision between the jet and the stationary interstellar gas (Source: J. Hester and NASA).

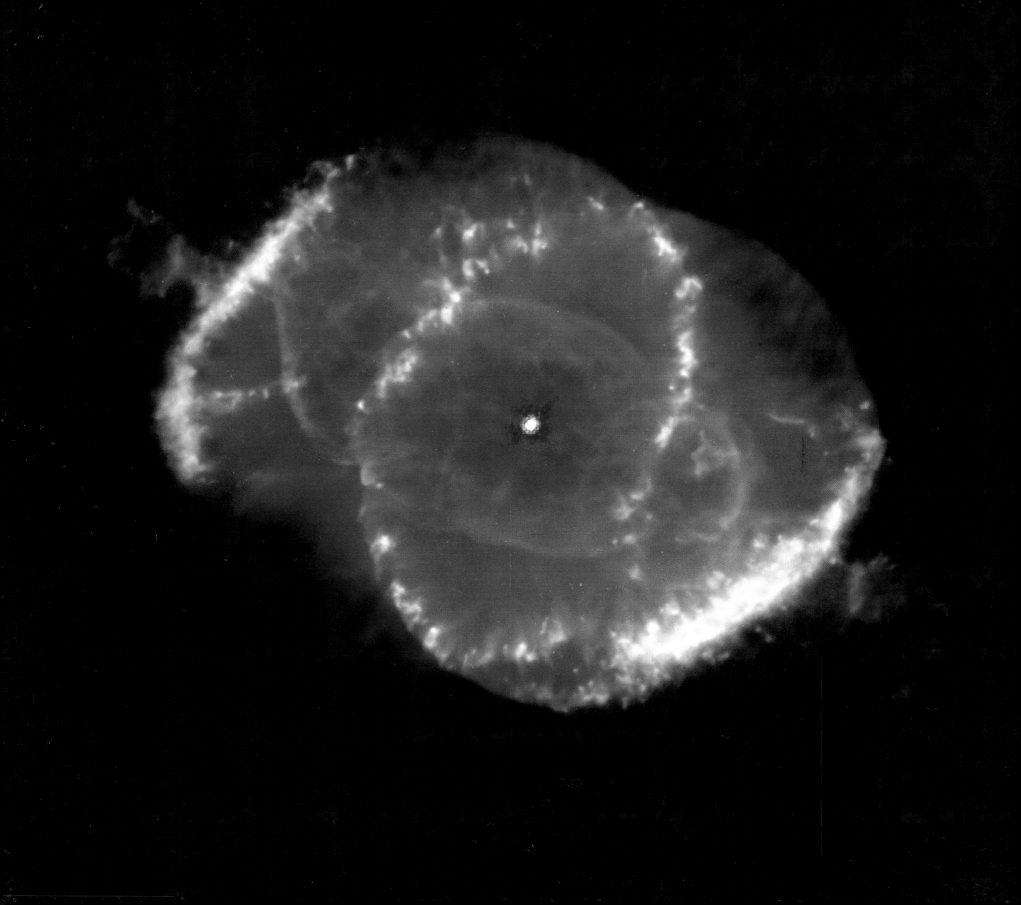

This planetary nebula called NGC 6543 shows off as a colorful (left) gas cloud – probably the remains of two dead stars (Source: J. Harrington, K. Borkowski, and NASA).

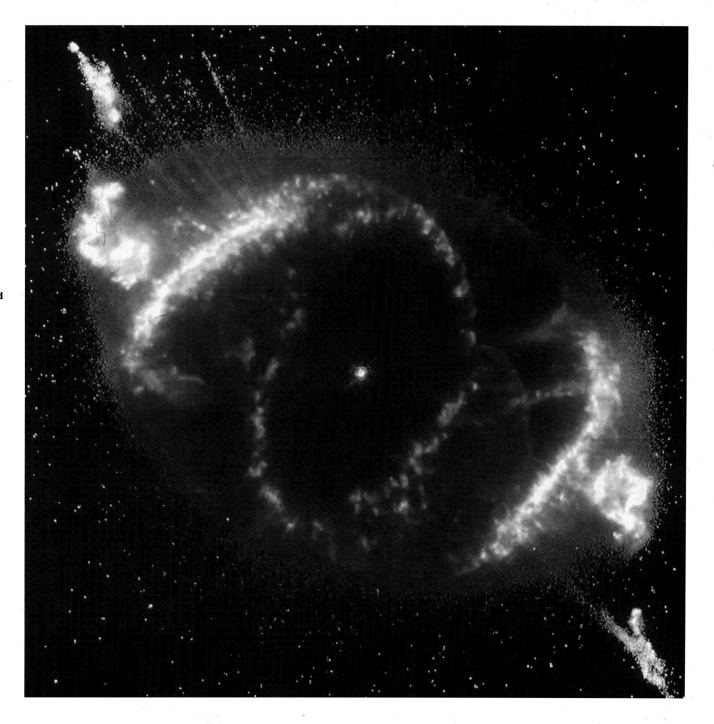

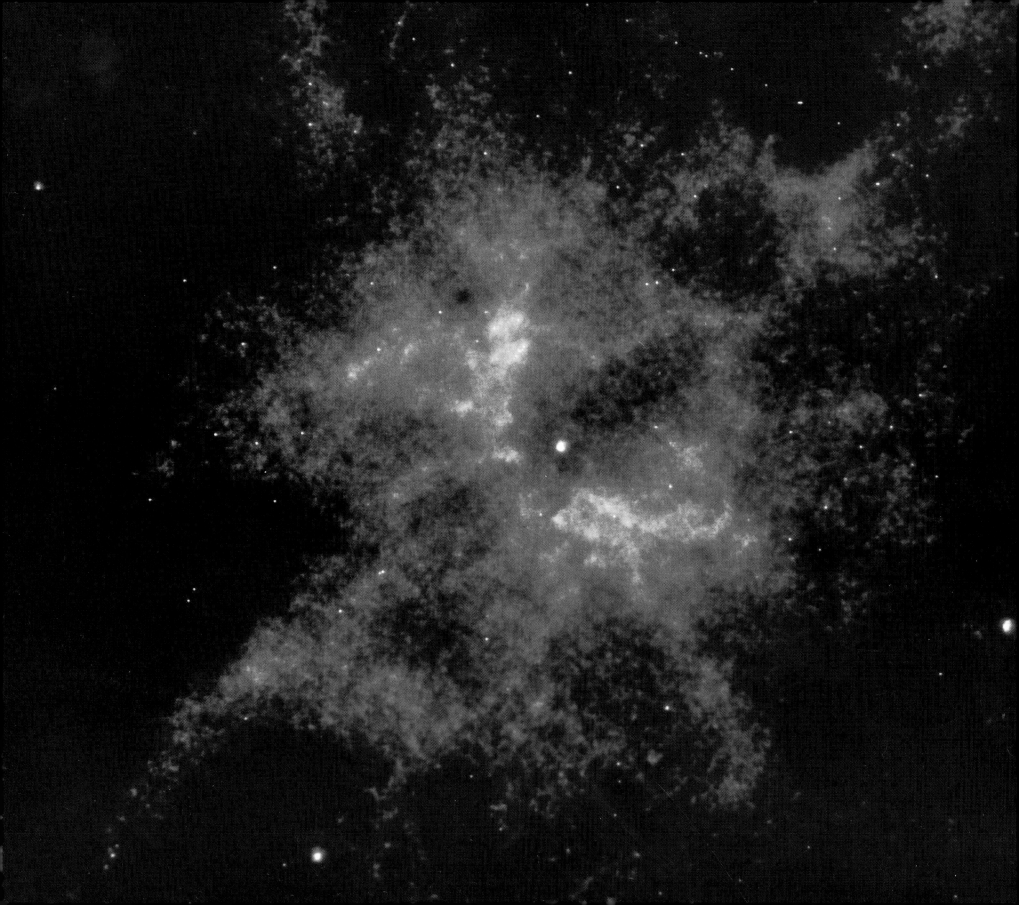

Left page:
Another planetary nebula – NGC 2440, where Hubble was able to show the central star distinct from the gas for the first time (Source: S. Heap and NASA).

Death of a great star: A giant with a mass much larger than our sun exploded, and illuminates bizarre rings. The outer rings of supernova 1987A are extremely thin (Source: C. Burrows and NASA).

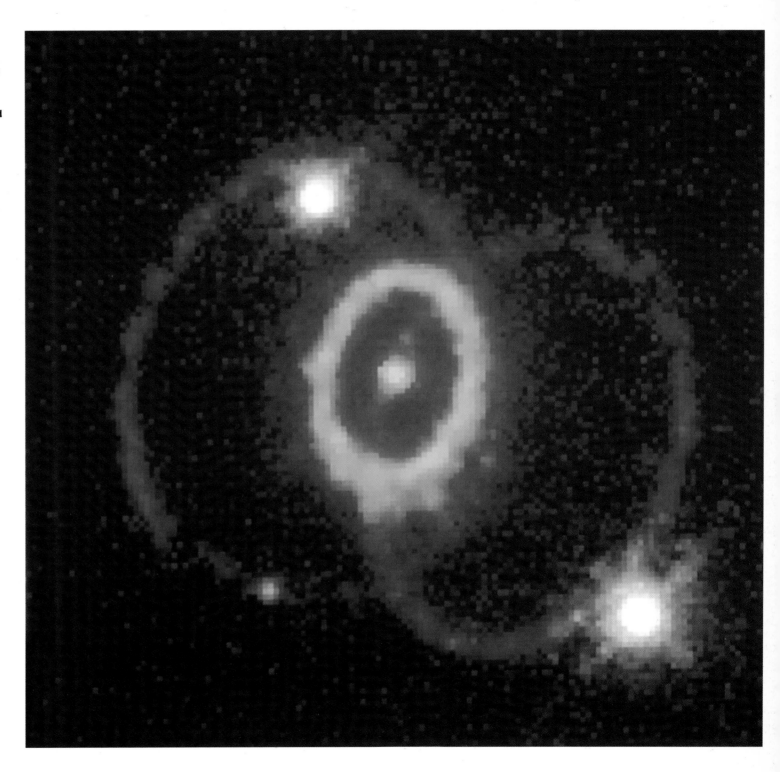

These filamentary nebulae are the remains of a star as well: The Cirrus nebula, as Hubble saw it before (left) and after (right) its repair. The colorful glow results from the collision of the gas ejected by the supernova explosion with the existing interstellar gas (Source: J. Hester and NASA).

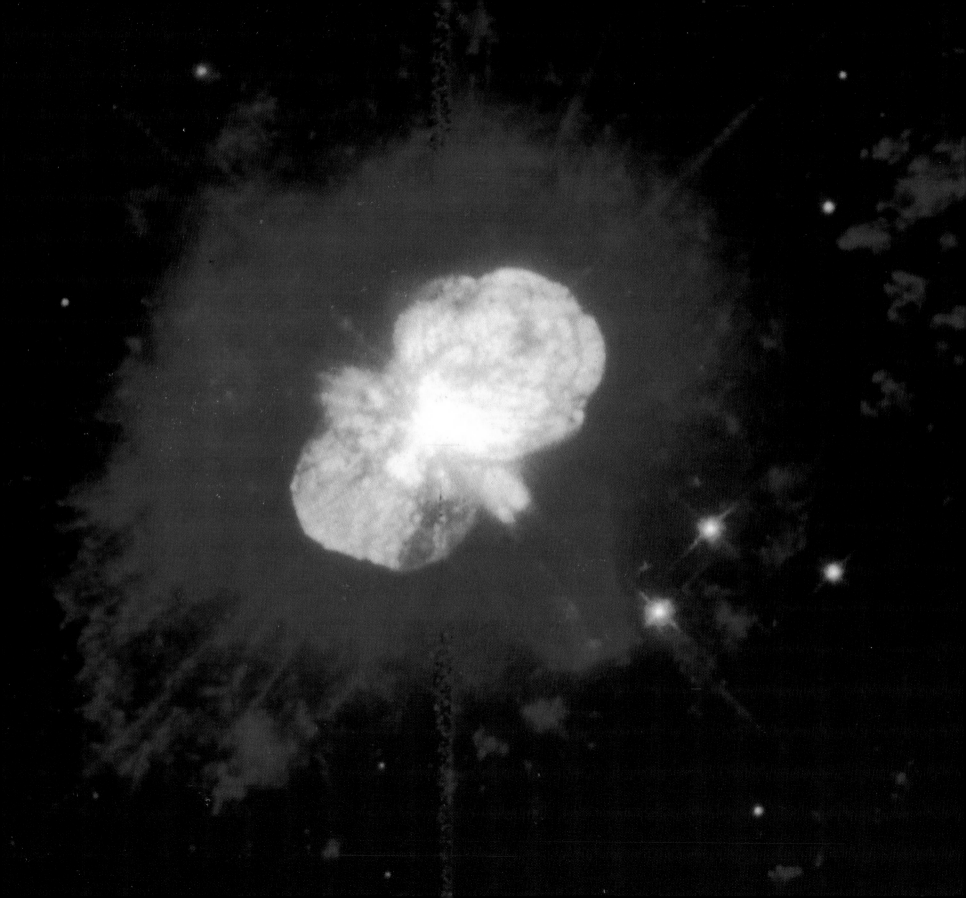

Nobody knows precisely why Eta Carinae changes its brightness, or why it sheds such a spectacular gas envelope (Sources: J. Hester and NASA, R. White).

lifetime. Low-mass stars like our own sun become "red giants" a short time before their death (however, this stage of stellar evolution is several billion years in the future for our sun), and finally eject a fairly spherical shell of thin gas. This shell is called a "planetary nebula," not because it has anything to do with planets, but because these compact nebulae looked to eighteenth and nineteenth century observers very much like images of distant planets. In the center of the nebula the dying remnant of the original star can often be seen: a hot white dwarf star, about the size of Earth, but almost as massive as the sun.

Hubble observed such a planetary nebula: NGC 6543 (pages 110–111), a small but bright object, discovered by William Herschel at the end of the eighteenth century. In 1864, William Huggins pointed a telescope equipped with a primitive spectroscope toward that nebula and recognized the spectrum of light-emitting gas. This was a great step toward an explanation of the nature of these nebulae – before that, several astronomers had claimed to have resolved these objects into individual stars!

Very detailed images of planetary nebulae had been taken with ground-based telescopes before, but Hubble was able to go one step further. Within the nebula, specific layers or envelopes of gas were identified, corresponding to a sequence of ejections of materials over time, and indications of high-velocity jets as well as condensations of gas caused by shock fronts were seen. A planetary nebula appears to be a much more turbulent place than we had thought.

It is unclear, but an interesting point of speculation, whether the complex structure seen in those images can be explained by a double star system inside the nebula.

Close-up of Stellar Death: Supernova 1987A

Planetary nebulae and white dwarfs are remains of stars with masses comparable to our sun. But if the original star is much larger, then it will go through various giant and supergiant phases, increasing its luminosity and using up its nuclear fuel faster and faster. Finally, no material to produce energy from nuclear fusion will be left: Its innermost core, now consisting mostly of iron, collapses into an extremely dense object, a neutron star or even a black hole depending on its mass. At the same time, the outer layers of the star are driven outward with high speed. In its death struggle, the star lights up one final time as a supernova.

Such a supernova appeared on February 21, 1987, in the Large Magellanic Cloud, a small galaxy neighboring the Milky Way and visible only in the Southern Hemisphere. Supernova 1987A, as it was called, being the first to be discovered in that year, reached an apparent brightness of 2nd magnitude, comparable to the stars in the Big Dipper, and then started to dim slowly. Now it is a very faint star, and in a few years it may be very difficult to observe. But this supernova left its characteristic calling card. The ejected gas expands with a speed of up to 10,000

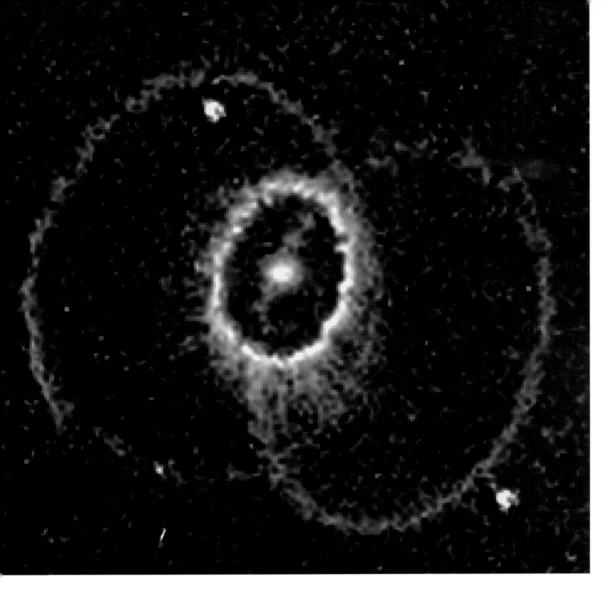

In addition to the color image on page 113, this computer-enhanced version shows the extreme thinness of the outer rings particularly well (Source: C. Burrows and NASA).

the stellar wind in earlier phases of the original star's evolution. One day, the gas cloud produced by the supernova explosion will catch up to this older envelope. Both will expand into space for millions of years, slowly mixing with interstellar matter and enriching it with heavy elements produced by nuclear fusion in the original star. From such star dust a new generation of stars may be formed. The universe, like human life, undergoes a continuous cycle of formation and fading away, and each dying star carries the seed of the next generation.

The Hubble image of Supernova 1987A (page 113) shows three rings. The small central ring as well as the two outer ones had been observed from the ground, but Hubble is capable of seeing them much more accurately – their explanation, however, remains unclear, as the standard model derived from ground-based observations had to be thrown out after Hubble's images. All three rings are tilted with respect to the line of sight, so that they appear to overlap; but they are positioned in three different planes. The smallest, brightest ring lies in a plane containing the location of the supernova, the other two are in front and behind it, respectively. Even at Hubble's resolution, the two outer rings do not have any noticeable thickness, posing a number of problems if they are physical bodies. But could they be some kind of optical illusion instead, projected on previously existing layers of interstellar matter? To generate a beam of light or particles to illuminate the outer rings, a very compact object – a neutron star

kilometers per second into space, forming a spherical envelope around the former star. Although Hubble was able to discern this envelope early on because it was slightly more extended than the image of a star of comparable brightness, it is too soon after the explosion to see any details clearly. But Hubble provided a good deal of related information. The energetic radiation released by the supernova as part of the collapse of its core illuminates preexisting gas in its vicinity. This older gas is part of a slowly expanding envelope, comparable to a planetary nebula found around other types of aging stars, and formed by

or even a black hole – with a close companion would have to exist in the center of the system. Material transferred from the companion star to the compact object would be heated up and ejected into space, together with energetic radiation. As the compact object rotates and precesses like a top under the influence of its companion star, these highly focused beams of radiation or particles could "paint" the two outer rings on layers of existing matter.

However, it is unclear why the origin of these symmetric beams does not coincide with the location of the supernova. Is there a close companion at a distance of about a third of a light year, having undergone a supernova explosion of its own a long time ago? Suffice it to say that the structure and origin of the three rings observed around Supernova 1987A poses an extremely difficult challenge to anyone attempting a complete and consistent explanation.

An interesting attempt at an explanation comes from England: Two stars encircling each other could have created the structure – not as an optical illusion, but as real structure in space. Stellar wind emanates from one of the stars (a red giant, later to become the supernova) and heats up as it passes the companion. The hot gas expands into the cooler remaining wind and forms two cones, which become rings as the red giant turns into a blue giant and develops a higher wind speed shortly before the explosion. Unfortunately however, this explanation is not perfect. This process would result in a shape of much higher symmetry than observed.

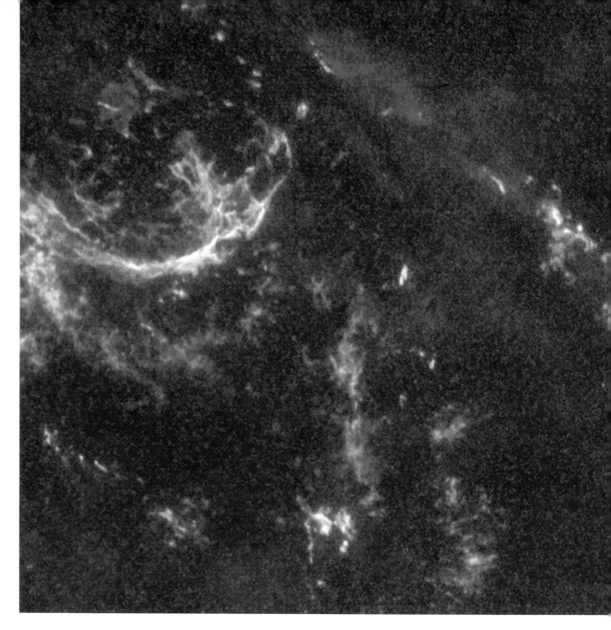

With the supernova itself fading each year, observations of the mysterious rings are becoming easier, and recent Hubble findings suggest that we may be able to measure the outer rings' thickness after all. Furthermore, faint luminous matter is now being detected between the two rings – are they perhaps the shells of two giant bubbles? An answer may come by 2000, when many of the faster ejecta of the 1987 explosion will probably hit the inner ring. At this point, according to one

An old supernova remnant in the Large Magellanic Cloud: N 132D. The blue-green filaments are oxygen-rich gas; they glow as they move through the reddish interstellar clouds – and Hubble can see all this over a distance of 169,000 light years (Source: J. Morse et al. and NASA).

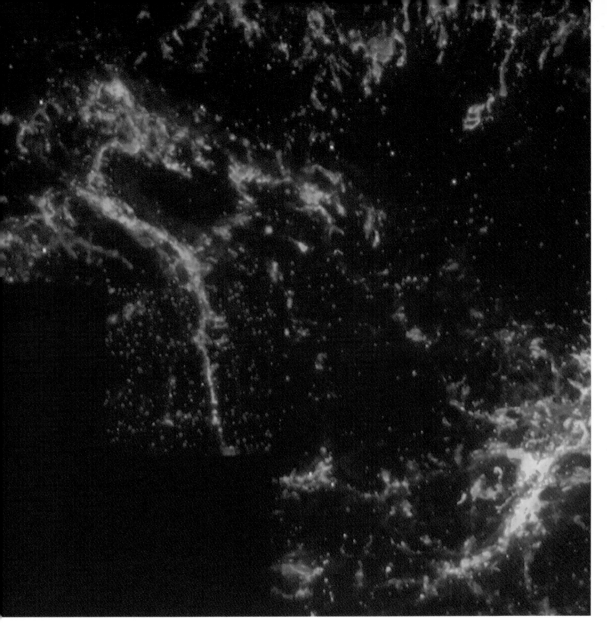

A small detail from the barely 1000-year-old supernova remnant M1: the Crab nebula in our own galaxy (Source: ST ScI).

theorist, "all hell breaks loose." Others speak of prodigious fireworks increasing the ring's brightness tremendously, illuminating the complex gas structure around the dead star.

A Stellar Ruin: The Cirrus Nebula

Supernovae explode in our own galaxy as well. The most recent ones were observed in 1572 and 1604, and it is quite possible that a supernova in our own cosmic home may happen within the next few decades. Supernovae are estimated to occur in our galaxy about every 10 years, but the light of most of them is absorbed by interstellar dust. Even in prehistoric times supernovae occurred: Pulsars are remnants of such events. These are rapidly rotating neutron stars, emitting a beam of radio waves that happens to point toward us during each rotation, producing a regular pattern of pulses. Only in rare cases can these objects be identified in visible light. They are the remains of the collapsed star that produced the supernova.

But supernovae leave behind much more. The gases ejected by the explosion, still moving through space with high velocity, can be observed in several cases. As they do so, they collide with interstellar gas clouds and heat them to high temperatures. These collisions produce X-rays and can also be observed in the radio and optical part of the spectrum. One of the most famous ones is the Cirrus nebula in the constellation Cygnus.

In 1784 the Cirrus nebula (NGC 6960/6962) was discovered by William Herschel using a 0.45-m reflecting telescope. This object is at a distance of about 2,600 light years and has a diameter of approximately 107 light years. It expands at about 80 kilometers per second: The supernova whose remains it represents ejected that material about 30,000 to 40,000 years ago (only yesterday in cosmic time scales) with a speed of several thousand kilometers per second. However, interaction with interstellar matter slowed it down over time to the current speed.

Hubble's images (pages 114–115) contain only small details of the Cirrus nebula; they show the structure of the glowing gas behind the expanding shock front generated by the supernova. This shock compresses and heats the interstellar matter. It is then capable of emitting visible light, providing us with information about its spatial distribution. The images – one taken before, one after Hubble's repair – show the shock running through dense clumps of matter about the size of our solar system. The bluish band of light is possibly a quantity of gas, ejected by the supernova, traveling through space at 1,400 kilometers per second. It will eventually catch up with the faster shock front, which has been slowed down by interstellar matter. Supernova remnants like the Cirrus nebula are fascinating large-scale laboratories for physics of shocks.

A Different Kind of Stellar Explosions: Novae

Stellar explosions that are less dramatic than supernovae occur much more frequently. Several dozen nova explosions take place in the Milky Way every year. But these are not catastrophic events like supernovae, only surface phenomena that do not effect the structure of the star.

Let us imagine a white dwarf, evolved from a red giant: a star the size of Earth but with the mass of the sun. Having used up all of its hydrogen, it cools slowly as a sphere of helium, carbon, and oxygen. After several hundred thousand years, it will have cooled so much that it is not visible any longer.

Stars do not only exist as singles, but also – and even more frequently – in double star systems. Configurations are possible where one star has evolved into a white dwarf, while the other star could be a red dwarf. Suppose these two stars are very close together, circling each other every few hours. The gravitational pull of the white dwarf is large enough to dislodge hydrogen from the outer layers of the red dwarf, forming a disk around the white dwarf. From this disk – the so-called accretion disk – material slowly rains down onto the surface of the white dwarf to create an ocean of hydrogen. But hydrogen can serve as nuclear fuel! Once a critical mass has assembled, and the density at the lower layers has grown to ten thousand times that of water, nuclear reactions set in abruptly, as in a hydrogen bomb, and eject the surface layer into space.

Such an explosion can be observed, and it is called nova, from the Latin *stella nova* meaning "new star" – a historical, but misleading name, as we now know, since the star is not newly formed, it only appeared that way to previous generations of astronomers. The expansion velocity of the ejected gas is about 1,000 km/s. From the ground, the envelope released by the nova can usually be observed after a few decades as a faint circular disk or ring around the now quiescent white dwarf.

The nova Cygni 1992 (named for the constellation Cygnus, or swan, in which it appeared) was discov-

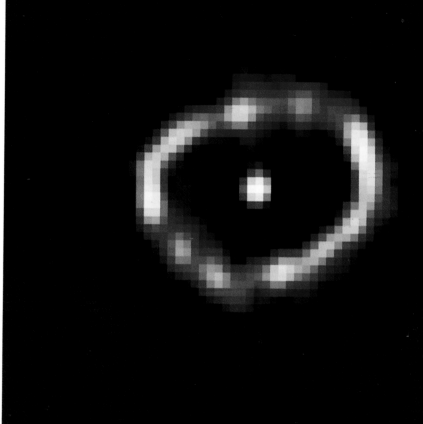

A "small" stellar explosion, the envelope of Nova Cygni 1992. Hubble can watch it expand, left on May 31, 1993, and right at the beginning of 1994. The ring has become more elongated over time (Source: F. Paresce, R. Jedrzejewski, and NASA/ESA).

ered in the early morning hours of February 19, 1992, by amateur astronomer Peter Collins of Boulder, Colorado. The development of the eruption was followed intensely with telescopes from the ground and with Hubble from space. Direct images of the expanding shell were taken by the Hubble on May 31, 1993, and in January 1994.

The image on the left, taken 467 days after the eruption, shows the typical disk-like appearance of the shell for the first time, and some kind of bar in addition. But because the Hubble was still suffering from spherical aberration at that time, details are not clearly visible. The image on the right, taken seven months later with corrective optics in place, shows an increase in size of the ring – its diameter grew from 126 billion to 163 billion kilometers. From the size of the ring and the expansion velocity at the time of eruption, the distance and absolute brightness of the

nova can be determined; this distance is about 10,000 light years.

Not only novae exhibit sudden bursts in intensity. The rare class of so-called luminous blue variables undergoes similar variations. These strange stars release matter into space at varying rates, but the reason for their instability is largely unknown. They are among the most massive stars in the universe. One of the prototypes for luminous blue variables is the star Eta Carinae in the southern sky.

Eta Carinae was cataloged as a star of 4th magnitude by astronomer Edmond Halley in 1677. By 1730, it had reached 2nd magnitude, then grew fainter, then brighter again. By 1827, Eta Carinae was a star of 1st magnitude, reaching its maximum in April 1843 – after Sirius, it was then the second-brightest star in the sky. The energy released by this mysterious star at that phase is comparable to that of a supernova, and this process has lasted for centuries. Later on, the brightness decreased down to 8th magnitude. In recent decades it has been slowly increasing. The absolute magnitude at the time of maximum brightness is hard to imagine – it corresponds to the magnitude of four million suns! We believe Eta Carinae is extremely massive – about 150 solar masses – and will end its development as a supernova in the not-too-distant future. Even today, the star is only borderline stable because of the tremendous amount of energy it generates, and it continuously releases matter into space.

Hubble observed Eta Carinae twice. The last image (page 116), taken after the servicing mission, clearly shows the gas clouds released over the last century. The red glow surrounding the star is produced by material ejected with a velocity of almost 1,000 kilometers per second. The white nebular structure in the vicinity of the star consists of two cones and is mostly made up of dust, reflecting the light of the star. It could easily accommodate our entire solar system!

In general, it is assumed that a massive equatorial disk funnels the released material toward the poles of the star, giving rise to the symmetrical structure. The variety and complexity of clouds and beams, as seen in the Eta Carinae images taken by the Hubble, will provide impetus to study these unstable stars for years to come.

Our Galactic Home – The Solar System

Compared with the vast worlds of galaxies and stars, our solar system is just an arrangement of small pebbles in space, and in comparison to even a very average star like our sun the planets of the solar system are almost nothing: 99 percent of the mass of the solar system is concentrated in the sun. Most of the rest is made up of the planet Jupiter, and some of Saturn, Uranus, and Neptune. Earth and her neighbors Mercury, Venus, and Mars – not to mention moons and asteroids – are mere dust specks.

In spite of this, every one of these planets harbors a world of its own. The more the solar system is explored by planetary probes and the more we know about the individual planets, the more fascinating our galactic home appears to be. Even in the close vicinity of Earth, a lot of questions remain to be answered, and new discoveries are waiting to be made. The Hubble Space Telescope has contributed to this area as well. Let us come home from the stars and begin an imaginary journey through the solar system, from the outside in. But before we start, we have to investigate a question that has involved the Space Telescope, at least indirectly.

Are We Alone in the Universe?

Looking at the stars on a clear night, everybody must have asked this question quite automatically, and since the earliest times humanity has pursued the question of other life in the universe, without a chance for much more than speculation. Attempts at an answer had been confined to philosophers and science fiction writers.

But for the last 20 years or so, science has begun to explore the question of extraterrestrial life. Since life as we know it cannot exist on or in stars with their high temperatures, but only on planets surrounding them, the most important task becomes the search for a star with a planetary system. Why shouldn't there be a sun among the billions of stars, surrounded by a planetary system similar to ours, where the development of life could be possible?

To prove that planets existed around other stars, and especially to obtain an image of them, became the Holy Grail of astronomy. With the discovery of dust disks around many young stars, and with simulations showing that planetary systems similar to our own almost had to evolve from them, confidence has been growing – but the basic problem in searching for distant planets had remained. Stars are so immensely bright compared with the reflected light coming from any orbiting planets that the latter would be invisible. Before launch it had been said that Hubble could overcome this problem, but specialists knowledgeable about its optical characteristics were skeptical. Even if the specifications were met, the residual roughness of the mirrors would scatter enough light from a star to make it very difficult to discern even a large planet close by. However, it was not widely known that Hubble had a chance to find other planets in a more indirect way. The Fine Guidance Sensors (FGSs), which orient the telescope in space with an accuracy of only a few thousands of an arcsecond (or milliarcseconds), are also capable of measuring distances between stars in their field of view with comparable precision. This indirect technique is still available, although in slightly degraded form because of Hubble's faulty main mirror.

Measurements of angular distances of stars with an accuracy in the milliarcsecond range were also performed by the European astrometry satellite Hipparcos, which systematically scanned the sky beginning in 1989 to establish a much more accurate reference frame of absolute stellar positions. But while Hipparcos observed stars in the entire sky down to a relatively modest brightness limit, Hubble's FGSs can measure much fainter stars in selected small areas of the sky. The FGS astrometry team had come up with a number of projects, ranging from distance measurements of certain stars to investigations of binary stars, and to the search for planets around other stars. How does this work? The principle is quite simple: A planet and its sun orbit about a common center of gravity. Proportional to their respective masses, this point is significantly closer to (or within) the star because of its much greater mass. At first glance one may even have the impression that the planet revolves around a completely fixed star, but more accurate measurements will reveal that the star moves in a small orbit around the center of gravity. This periodic movement could be measured by the FGSs, at least for relatively close stars, *if* they had planets the size of Jupiter.

Astrometry with the FGSs is a difficult task, and only a modest amount of observing time is available because of all the other demands on observing time with the Hubble. But there is a growing database of measurements, particularly for the star closest to our sun, Proxima Centauri. Until mid-1995, there had been 48 series of measurements of its angular separation from several other stars, each accurate to about 0.002 arcseconds; variations of only one thousandth of an arcsecond should have been noticeable, but no planets – at least up to a limit of half a Jupiter mass – have been identified around this star as yet.

Can we expect the Hubble to provide the sensational discovery of a planet around another star? It may be a race between the most powerful telescopes on the ground and the Hubble. Observations already taken by both could be interpreted as containing images of planets around other suns. It is probably only a question of time before we have a confirmed discovery.

Where Are the Limits of Our Solar System?

Let us return to our imaginary tour through the solar system. When would we cross the outer boundary of the system? This question appeared to be answered definitively after the discovery of the ninth planet, Pluto, more than 60 years ago, but today it may be more difficult to decide. When the Space Telescope was launched in 1990, our solar system ended with the well-known planets Neptune and Pluto, which take turns in being the most distant from the sun. Only speculation existed about the world beyond: Perhaps there is a gigantic collection of icy objects a few kilometers in diameter, which could be diverted toward the inner reaches of the solar system on rare occasions to become comets, but there was no chance to observe them in their natural habitat. However, in 1992 a surprising discovery was made with a telescope in Hawaii: a body of several hundred kilometers' diameter was found outside the orbits of Pluto and Neptune, and a second one was discovered in 1993. By 1995, calculations predicted the existence of 30,000 to 50,000 such "trans-Neptunian objects" with a diameter of more than 100 kilometers, and at distances of between 4.5 and 7.5 billion kilometers from the sun. They represent a completely new component of the solar system, and probably are the bigger brothers of the comet cores in their holding area – there may exist about 1 to 10 billion of the latter down to a diameter of 1 kilometer. The hypothetical "Kuiper belt" of icy bodies outside Neptune suddenly became reality. A race for new discoveries began, and therefore a Kuiper belt search team for Hubble was formed as well.

By April 1995 the first success could be announced: About 59 candidates of Kuiper belt objects had been found by Hubble – albeit at the limits of its performance. The identical field had been observed 34 times with an exposure time of ten minutes each, and the images had been superimposed accurately: Stars and galaxies were identified and eliminated from fur-ther studies. Then the images were co-added again, but this time with slight offsets, in such a way that a typical Kuiper-belt object would remain stationary. A total of 44 realistic trajectories were tested. As a control, the experiment was repeated with 44 unrealistic trajectories, and then the co-added images were analyzed for potential objects. The result: 244 candidates for the "realistic" offsets, versus 185 for the unrealistic ones. This yielded a significant excess of probably real objects with brightnesses around 28th magnitude. The core of comet Halley would be about that faint at a comparable distance from the sun. The claimed detection of small Kuiper Belt objects has met with considerable skepticism, however, from those who think Hubble's current instrumentation is just not powerful enough to yield convincing images of a moving 28th-magnitude object such as a comet core at that distance. However, a second set of comet cores at a much larger distance from the sun, the famous Oort cloud, will remain permanently inaccessible to Hubble (and to any other conceivable telescope).

Pluto

Pluto, the outermost planet of our solar system, was discovered in 1930. That late date is not surprising given its average distance from the sun of 5.95 billion kilometers. Nowadays, its status as a true planet is sometimes questioned. After the discovery of Kuiper belt objects by Hubble, Pluto could be counted as their largest representative,

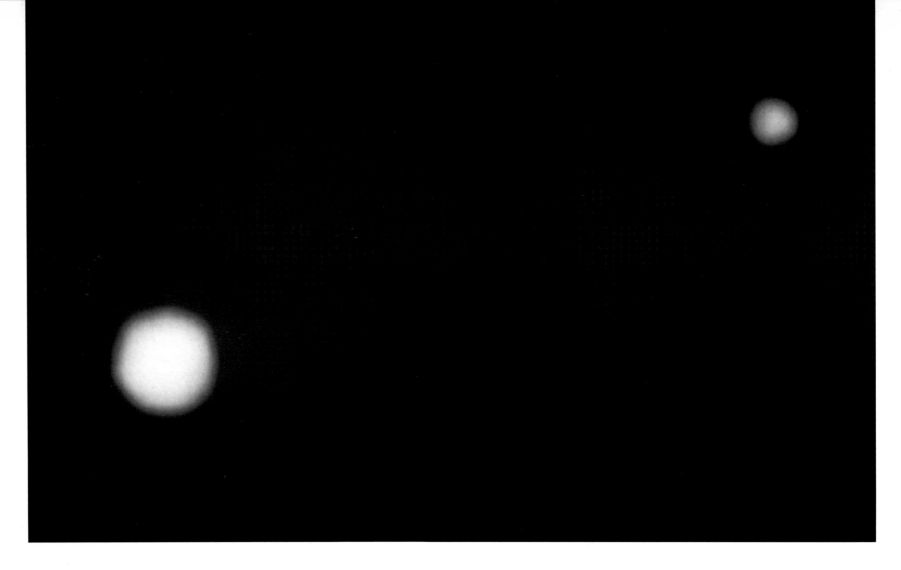

but its discoverer Clyde Tombaugh and his friends are fighting for its recognition as a true planet. It is actually a double planet, since Pluto's moon Charon has half its diameter: such a pair is unique in the solar system. Space probes have not yet visited the strange pair, and therefore the most detailed images of the system are due to Hubble. In 1990 it succeeded in showing the two bodies fully resolved. On ground-based telescopes, Charon shows up only as a faint extension of the image of Pluto. After the installation of corrective optics for Hubble, even better images were obtained; bright and dark regions became visible on Pluto's surface, but not enough information is available to determine their

detailed nature. Surprising facts can be deduced from Hubble's detailed measurement of these two bodies' movements. Hubble's data showed that Charon is eleven times further away from their common center of gravity than Pluto, which therefore must have eleven times Charon's mass.

Up to that point it was only known, from their distance and orbital period, that both of them together contained about 1/400 of the mass of Earth. Later on, the aspect of their orbit happened to allow observers to witness a number of mutual occultations, so that their diameters could be determined to be 2,320 and 1,270 km, respectively, as confirmed by Hubble to an accuracy of 1 percent. Now the equations can

Pluto and its moon Charon – more a double planet than a planet and its moon. The two had not been seen so clearly before. Over 4.4 billion kilometers Hubble has fully separated their images for the first time.

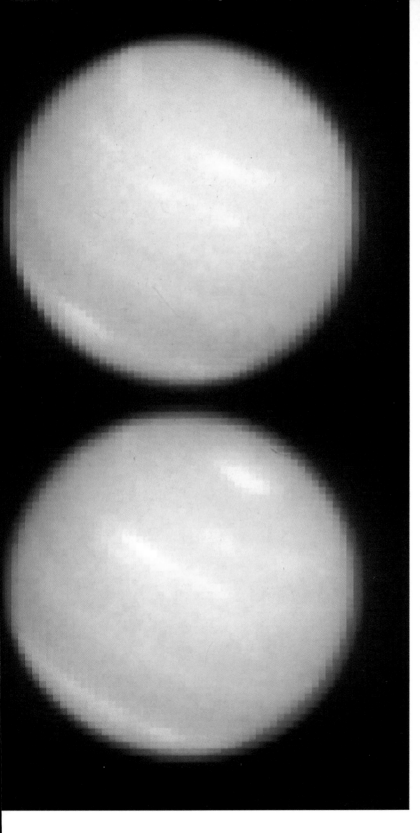

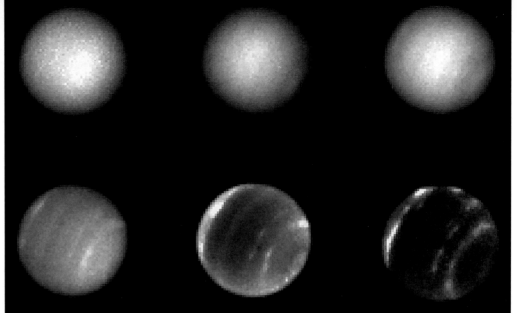

This is how Hubble saw planet Neptune in June 1994. The black-and-white image, showing Neptune's clouds with great contrast, was taken at a wavelength of 889 nm, where methane in deeper cloud layers absorbs the majority of light, so that the higher clouds stand out clearly (Source: D. Crisp, H. Hammel, and NASA).

be solved, yielding a density for Pluto of 2.1 g/cm³, and for Charon of only 1.4 g/cm³. Therefore, Pluto must contain a significant fraction of rocky materials, whereas Charon is largely a big snowball. This fits with a popular theory for the origin of this rare double planet, according to which the system would have been formed by a collision of a proto-Charon with a much more icy proto-Pluto. The latter would have lost most of its layers of ice, while Charon remained an icy sphere. This scenario is also supported by the different spectra of the two bodies, and by their colors. Charon appears much bluer than Pluto, as Hubble determined by images of the pair through different filters.

Neptune

A world completely different from solid Pluto awaits us at the next stop of our journey toward the sun: Neptune. Over an average distance of 4.5 billion kilometers, this distant gas giant made up largely of hydrogen and helium, with a diameter of 49,420 km (approximately four times that of Earth), has an angular diameter of only 2.3 arcseconds, just a diffuse sphere for ground-based telescopes. For Hubble, however, it is a dynamic world of clouds. During August 1989, the planetary probe *Voyager 2* encountered Neptune at close range and saw numerous bands of clouds and dark storms accompanied by bright clouds at high altitudes. After its repair, Hubble observed Neptune with increased resolution during the summer and fall of 1994, and generated surprise among astronomers. Many of the pronounced dark spots in the 1989 images had vanished, and new cloud formations had appeared. Even between June and November 1994 significant changes could be seen, which had not been suspected on a planet thirty times as far from the sun as Earth.

The dynamic changes of its atmosphere may be related to a strong internal heat source, retained by the planet over billions of years. For atmospheric scientists Neptune represents an interesting object in any case. And thanks to Hubble, further changes on its surface can be monitored for years to come.

Uranus

The next stop on our journey is Uranus, another gaseous planet, similar in composition to Jupiter. Its mean distance from the sun is 2.8 billion kilometers. When *Voyager 2* visited Uranus in 1986, it looked fairly dull. A few cloud formations could be made visible using sophisticated image reconstruction techniques, but other than that it appeared as a uniform green gaseous sphere; however, it was surrounded by an interesting system of thin rings. In addition, *Voyager* discovered several small moons of Uranus, and when

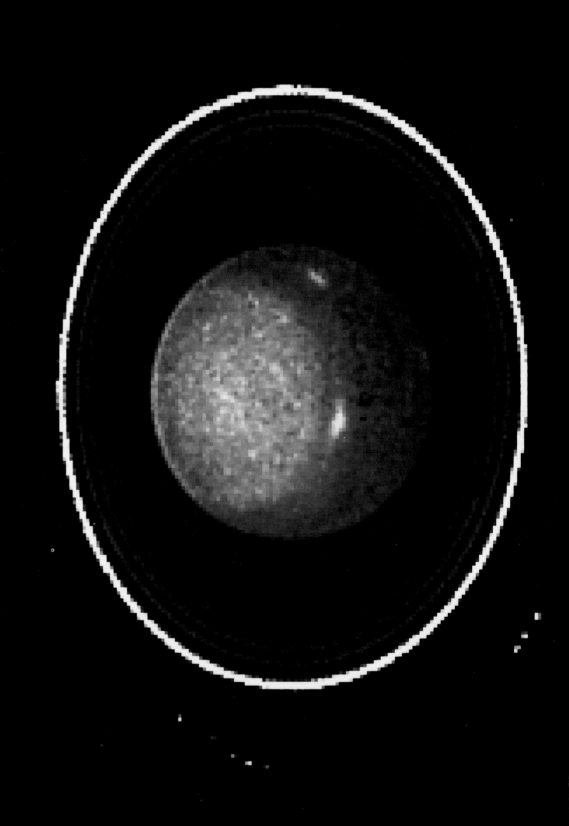

Uranus from a distance of 2.8 billion kilometers during August 1994. Several of the 11 rings are visible, as are the small moons Cressida, Juliet, and Portia (Source: K. Seidelman and NASA).

Hubble took images in August 1994 to determine better parameters for their orbits, it delivered a surprise: Uranus itself had something to show! The images revealed several individual clouds, with much better contrast than anything available from the 1986 visit. The difference may be due to the changes in illumination over eight years. Uranus' rotation axis is not approximately perpendicular to the plane of its orbit around the sun, as is the case for the other planets, but it lies practically in the orbital plane. In 1986 the planet was positioned in a way that placed the sun directly over one of its poles, so that the illuminated hemisphere received the same dose of light day in and day out. But by 1994 the sun appeared at about 55 degrees latitude over Uranus, so that for a given point in the atmosphere the position of the sun changed significantly over the course of a Uranus day. These changes could have induced interesting meteorological effects. Uranus does not have an internal heat source like Neptune's. It is fully dependent on the sun.

Saturn

Pluto, Neptune, and Uranus were discovered only after the invention of the telescope. But Saturn is the second-largest planet of the solar system, and in spite of its distance of 1.4 billion kilometers from the sun it is bright enough to be easily distinguishable from the fixed stars. It has been observed as a "wandering star" since ancient times.

Its most remarkable characteristic, however, can be seen only through a telescope: its spectacular system of rings, far overshadowing the ring systems of other planets in brightness and complexity. Therefore, it was possible for the Hubble, even with uncorrected optics, to provide dramatic images of the ringed planet during the summer of 1990 (see also page 50). These images showed several gaps between rings, the second widest of which, the Encke division, had never been revealed except from planetary probes. Hubble delivered a command performance during the fall of 1990, when a giant storm appeared in the cloud structure of the planet. From one day to the next, a bright white spot appeared and was expanded by complex winds over the following weeks. With some effort it was possible to reprogram Hubble, then still fairly error-prone, to systematically monitor the storm.

The results were summarized in a much-admired time-lapse movie. The individual images show the development of the white cloud with much better resolution than possible from the ground. A second turbulent event occurred on Saturn in December 1994, exactly one year after the installation of the second Wide-Field Planetary Camera. The events of 1990, which by now were fairly well understood, were repeated on a somewhat smaller scale. The phenomenon could be compared to a terrestrial thunderstorm. A warm packet of atmosphere begins to rise from the lower layers, breaks through the regular bands of clouds, and then begins to cool again.

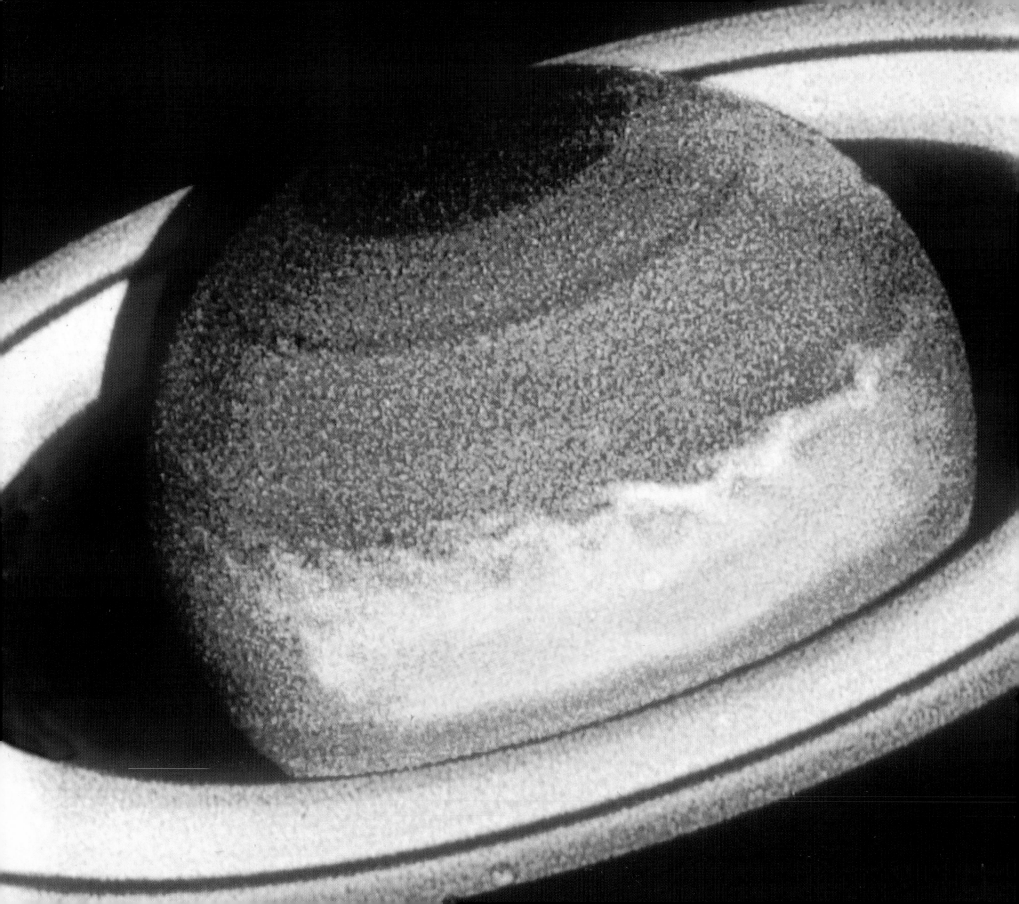

Storms on Saturn: The first, much larger storm appeared in 1990 (left), the second one at the end of 1994 (Source: R. Beebe et al. and NASA).

May 1995

Dione

Tethys

01:50:20.55 UT

Pandora

Janus

Rhea

06:24:20.55 UT

Janus

Enceladus

Rhea

08:02:20.55 UT

Earth crossed the plane of Saturn's rings on May 22, 1995. Shortly before, the normally majestic ring system appears only as a thin line (top); for Hubble, they remain visible even at the exact time of the crossing, as the sequence of three images (bottom) shows. In the windows, the contrast is dramatically enhanced, so that the rings and several small moons become visible. Janus, for instance, can be observed only on such occasions (Source: A. Bosh and NASA).

Facing page:
The surface of Saturn's moon Titan. Many of Hubble's images have been combined into a globe, here shown from four sides (Source: P. Smith, M. Lemmon, and NASA).

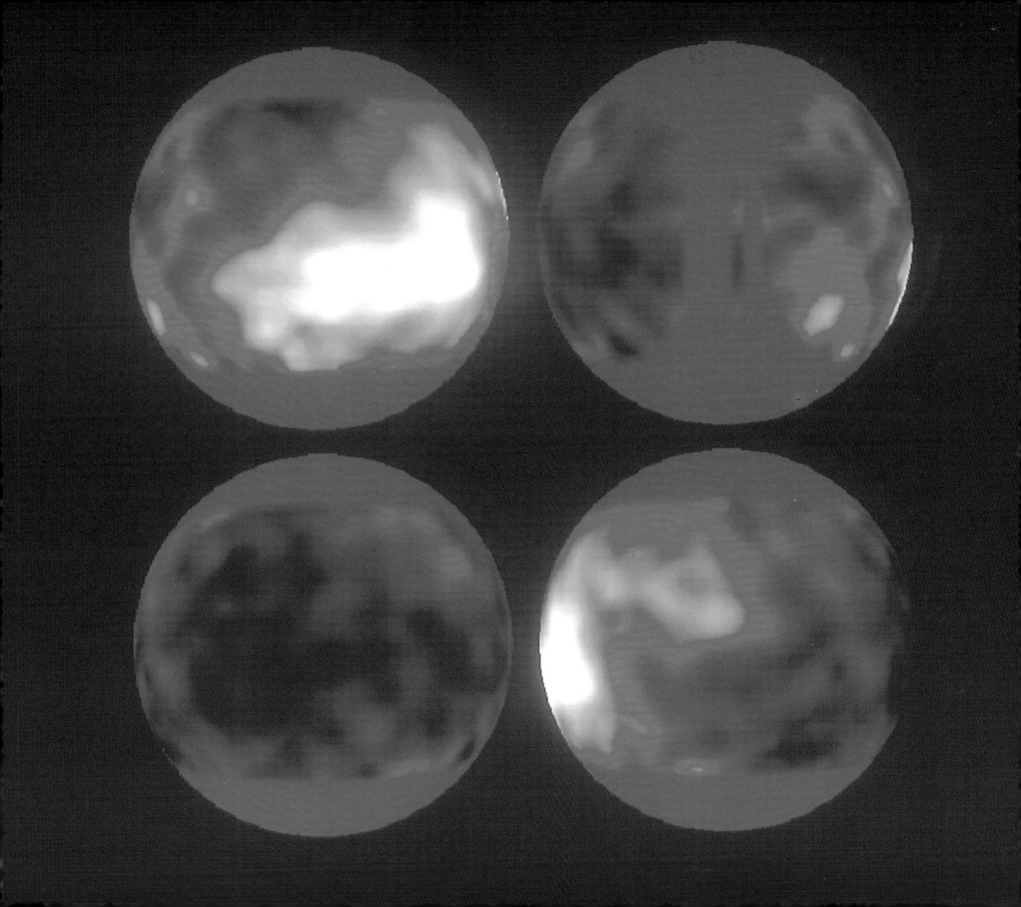

Ammonia starts to condense and forms bright white crystals, forming the white spot, visible even with a small telescope from Earth, over a billion kilometers away. But only Hubble can observe the shape of the cloud in detail. The wedge-shaped structure, which it assumed after quarter of a year, gives clues to the varying speed of Saturn's winds in different latitudes.

Approximately every fifteen years Earth and Saturn are aligned in such a way that Saturn's ring system appears exactly edge-on. For small telescopes on the ground, the rings simply disappear, but for Hubble these edge-on events are much more interesting. On May 22, 1995 such an event occurred. The space telescope took images in rapid succession of the rings, which appeared only as a thin line. Even at the moment of passing through the plane of the rings, they did not disappear completely. Despite being only several dozen meters thick, the edge of the rings reflects some sunlight. These events have scientific importance: The characteristics of the rings themselves can be studied well edge-on, small moons within the ring system or close to its border can be detected, and finally the view of Saturn as a whole is not disturbed by the rings, which normally cover part of the planet.

The ring plane crossings of May and August, as well as the transit of the sun across the ring plane in November 1995, have led to numerous discoveries about the Saturnian ring system. Hubble detected seven mysterious objects in orbit around the planet, only two of which could later be identified with known moons. One of the remaining five turned out to be an actual new moon of Saturn; the others are probably not solid objects at all but temporary condensations in Saturn's amazing F ring. This very narrow ring, just outside the main ring system, turns out to be warped in more than one dimension and is one of the most variable rings in the solar system. During the ring plane crossings Hubble was also able to image the G ring and the diffuse E ring, which lies far outside the usually visible bright rings. This work will help in the mission planning for Cassini, the probe that is due to launch in 1997 and enter the Saturn system in 2004.

Saturn is not only famous for its ring system, but also for being circled by a moon of the size of a small planet – it is larger than Mercury! Moreover, this giant moon is the only one with a prominent and stable atmosphere. This object, rightfully called Titan, disappointed the fans of spectacular images from planetary probes, when Voyager 1 had a close encounter in 1980. The clouds appeared monotonous and opaque. Titan's atmosphere was quite interesting from a chemical point of view, exhibiting similarities to the young Earth (it consists largely of nitrogen, like our own), but only vague speculations could be made about the body underneath. It was largely assumed that it would be covered by an ocean of liquid methane, because this gas appears in the atmosphere and must be constantly replenished – but then highly focused radar pulses, sent to Titan from Earth, showed that at least part of the moon's surface must be solid. Finally, it was determined that the atmosphere is transparent at certain wavelengths in the near infrared, which are accessible to the Hubble, so that it could see Titan's surface.

During October 1994 Titan's surface was systematically imaged by Hubble, and a complete map of the moon was generated by computer. Large parts of Titan appear dark, but there is a remarkably bright area about the size of Australia, the nature of which is the subject of intense discussions. Is it a continent surrounded by a huge ocean? This seems impossible, as the friction between ocean and solid land caused by tidal action would dissipate much of Titan's orbital energy, so that a circular orbit around Saturn would result; however, its orbit is elliptical. From the point of view of celestial mechanics, Titan could only have several small lakes without connections to each other (because then the tidal friction is much less), but how does this fit with the map derived from Hubble observations? One fact remains: When the planned planetary probe Cassini releases the camera-equipped Huygens capsule through Titan's atmosphere down to its surface at the beginning of the twenty-first century, it will find a more interesting and enigmatic moon, than had been expected after Voyager's visit more than twenty years earlier.

Taking Aim at Jupiter

This giant of gas beckons for a visit very strongly, as it has produced headlines all over the world during 1994. The largest and most interesting of the four gaseous planets is twice as massive as all the other planets combined (its mass is 318 times that of Earth). But compared to the sun, from which it has an average distance of 778 million kilometers, it is a tiny dwarf. The multitude of phenomena and colors in its atmosphere are unrivaled by those on Saturn, Uranus, or Neptune. In spite of its flawed optics, Hubble began very early to monitor the changes, and after the repair the images have reached the quality of those taken from planetary probes in close proximity to the planet. Jupiter by itself was certainly an interesting object for investigation by Hubble – but then a comet was discovered in the spring of 1993 that could impact Jupiter, as was surmised for the first time in May 1993. The comet was called P/Shoemaker-Levy 9 after its discoverers (the P stands for periodic or recurring comet, according to its trajectory, and the number 9 indicates that it was the ninth comet of that name), and its unusual appearance alone justified direct observations with Hubble. After a close encounter with Jupiter in 1992 it had broken up into about 21 fragments, now looking like a string of pearls on their orbit through space. The images taken with Hubble would help determine the sizes of the various fragments, which would in turn greatly affect any potential collision. But the first images from the summer of 1993 could not provide an accurate answer.

About at the time of the first servicing mission it became clear that all fragments of Shoemaker-Levy would impact Jupiter in July 1994, with speeds of

Comet P/Shoemaker-Levy 9 approaching Jupiter – a long "string of pearls," or comet fragments. Each consists of a core and a tail, made up mostly of dust (Source: H. Weaver, T. Smith, and NASA).

The giant planet Jupiter, awaiting the impacts of comet Shoemaker-Levy 9 in May 1994 (Source: H. Hammel and NASA).

about 60 kilometers per second. More and more frequently, Hubble images of the increasingly extended "string of pearls" were taken, which no longer even fit into a single frame of the camera. Each fragment had developed into a little comet of its own, with a nucleus, from which the sun evaporated some material to form a core region and a tail, which was forced away from the sun by radiation pressure. Around the cores of the little comets the dust appeared to be more dense, forming spherical layers around the centers – a good indication that solid nuclei existed and were still releasing dust. However, it turned out to be impossible for Hubble to see the nuclei clearly and to determine their diameters accurately: The fragments could be several kilometers in diameter at most, and the predictions for the effect of their impact on Jupiter varied from spectacular to unmeasurable.

This event of the century for observational astronomy was reason enough to schedule all sorts of observations on the Hubble, and on most observatories around the world, during the time of the impacts from July 16 through 22, 1994, and for the following weeks. Several large special projects had been prepared, and great hopes were pinned on Hubble's cameras. During the impacts, which by now had been calculated to better than minutes, images of Jupiter's edge would be taken, as the actual sites of the impacts were on the far side of Jupiter, just behind the edge.

But model calculations had indicated that a rising gas plume might be visible above the horizon before the actual impact site rotated into view.

As much observing time as possible (and necessary) was made available on the Hubble for the impacts. However, data transmission and calibration take some time. After the impact of fragment A, reports and images from ground-based observatories were already circling the world, when the first Hubble images appeared on the computer screens at the Space Telescope Science Institute in Baltimore. And indeed, the first ground-based as well as the first Hubble images showed a gas plume rising over the edge of Jupiter! Minutes later, the impact site had rotated into view, and the images now showed an impressive dark spot surrounded by a crescent-shaped halo.

During the following days, over 400 images were taken of the impacts or impact sites of the more than 20 fragments. Particularly impressive were the images taken in the ultraviolet, showing the dark spots left by impacts even more dramatically than in visible light. In the infrared, however, Jupiter appears quite dark, due to absorption of sunlight by methane in its atmosphere, but the impact sites were bright, proving that they were high in the atmosphere of the planet.

Hubble made its most spectacular observations during the impact of the so-called G-fragment. Normally, the exact pointing of the telescope is planned several weeks in advance. However, because of the uncertainty in the time and location of the impact, the exact position of the dark spot that was expected to result from the G-fragment impact was not known. Therefore, a real-time position update had to be performed, where an image of the area in question is taken, the coordinates measured, and several minutes later more accurate coordinates are uplinked to the telescope. This approach, which had been used before in similar situations, worked as expected, so that an hour and a half after the impact one of Hubble's spectrographs was able to take a spectrum of the exact location of the G-fragment impact. The data obtained in this observation are of great importance for the interpretation of the chemical processes in the cooling gas plume of the explosion resulting from the impact. Sulfur, carbon-sulfur compounds, and ammonia were identified from the spectrum, giving important data points for modeling the chemical processes.

Hubble was able to observe four impacts very closely. In addition to the A- and G-fragment impacts, data from the impacts of the E-fragment and from the last, the W-fragment, were taken as well. In all cases, the sequence of events was similar. First, a roughly spherical gas plume appeared, which then flattened out into the upper atmosphere of Jupiter becoming a typical dark spot as it rotated into view. The dark material must have formed very close to the time of impact, as the gas plumes were already darker than Jupiter's atmosphere when they became

visible over the horizon; only against the blackness of space did they appear bright. The clouds probably consisted of organic molecules formed in the cooling plume after the impact. Close inspection of the G and W impacts also revealed that at the time of the explosion – which happened just behind the horizon, on the far side of Jupiter – as well as several minutes later a faint glow could be seen right next to Jupiter's edge. These phenomena could be understood only after comparison with ground-based observations. It appeared that Hubble had succeeded (at least with the G impact) in seeing the fragments as they entered Jupiter's atmosphere. Like meteorites on Earth, the fragments began to glow as they became heated by friction in the upper layers of the atmosphere. On several occasions Hubble was able to see the glowing fireball of the explosion as it rose over Jupiter's horizon.

Perhaps even more surprising than the effects visible during the impacts – even on Jupiter's far side – was the longevity of the impact traces. Wisely, the Hubble team had obtained some observing time for the days and weeks following the impacts. So it was possible to follow the development of the impact sites as they changed from well-defined dark spots into structures affected by the winds on Jupiter, and then into smeared-out bands across the limb of the planet. This happened so quickly that the Hubble images by themselves were not sufficient to track the individual spots and structures. Only in comparison with the many ground-based images will

The most spectacular impact occurred from comet fragment G on July 18, 1994. The sequence shows the cloud from the explosion (bottom), and then the impact site as it rotates into view (Source: H. Hammel and NASA).

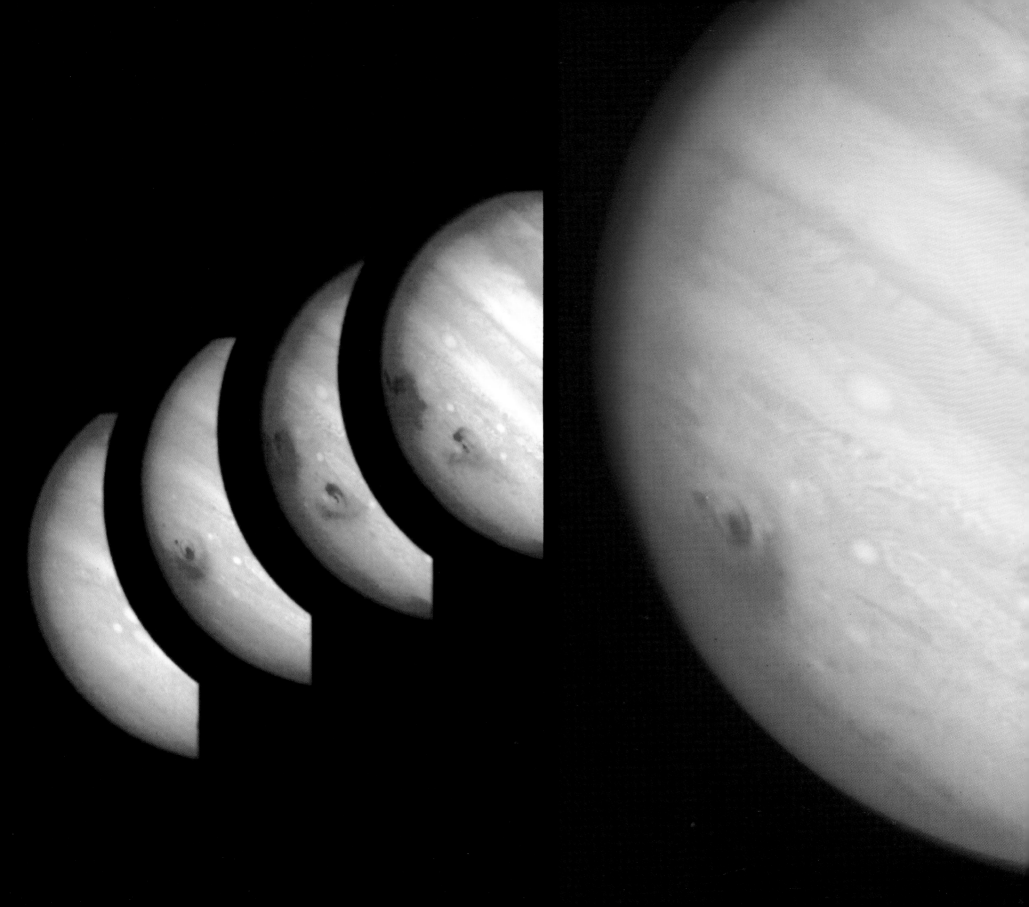

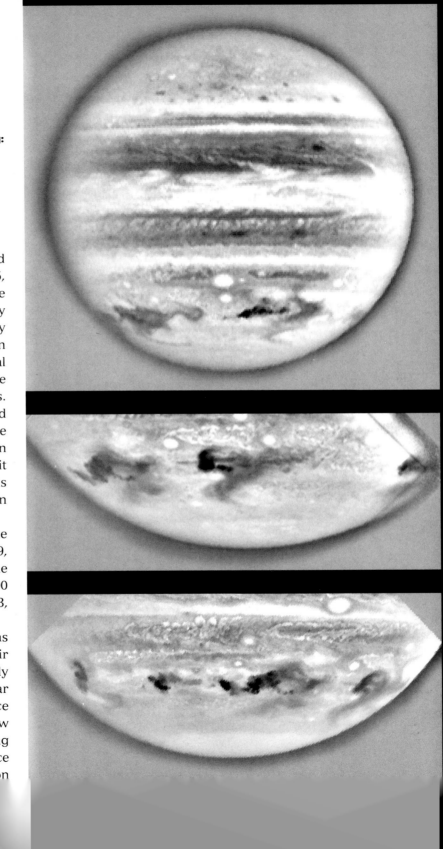

it be possible to identify the details of the winds and currents. Hubble observations ended on August 25, 1994, when the planet came too close to the sun to be observed safely, but they were resumed in February 1995. The conspicuous dark band had practically vanished. Several cloud formations could have been either impact remnants or normal meteorological phenomena. However, Hubble was still able to see the impact sites at ultraviolet and infrared wavelengths. At this point, Hubble observations of Jupiter had reverted to their normal frequency, now laying the foundation for the arrival of the Galileo probe in December 1995. Its field of view from the close orbit around Jupiter will be very small. Therefore, it is necessary to select the more interesting areas in advance, based on Hubble images.

In its observations of Jupiter, in particular on the occasion of the impact of comet P/Shoemaker-Levy 9, Hubble had done a great job. The Space Telescope, the object of great expectations before its launch in 1990 and of great controversy until its repair in late 1993, had delivered a masterpiece.

Let us now take a little detour to the large moons of Jupiter; for Hubble, they are small worlds in their own right. The most interesting moon is undoubtedly Io, volcanically the most active body in the solar system, continuously reshaping its surface. Since planetary probes have visited this moon only a few times, Hubble's capacity for long-term monitoring becomes important. Before its repair, the Space Telescope was already able to see a lot of detail on

Jupiter seven months after the impacts: at least in the visual range of the spectrum, the impact marks have become too faint to be visible, except for a few spots (Source: R. Beebe and NASA).

Io's surface. In visual light, Io looked identical to the images taken in 1979 by the Voyager probes, but significant differences show up in ultraviolet light. Areas appearing bright in the visual range appear dark in the ultraviolet, probably due to a thin layer of sulfur-dioxide frost, which absorbs UV but reflects visual light. Images taken after Hubble's repair even show individual volcanoes discovered by Voyager, for

instance Pele. And since Hubble has a larger selection of filters available, it became clear only now that the ejecta from this volcano have an unusual composition and are probably rich in sodium. Hubble made an interesting discovery on Europa, another moon of Jupiter, as well: it exhibits an extremely thin oxygen atmosphere, released in complex processes from its icy surface.

Jupiter and its moon Io –
a comparison of size (top).
Io images from March 1992
(center row) and from *Voyager*
(bottom right), taken in 1979,
and artificially reduced to
Hubble's resolution (bottom
left). The latter one can be
compared with the center left
one above it. The volcanic
landscape has hardly changed.
A Hubble image taken in
the ultraviolet (center right)
shows differences: areas that
are bright in the visible range
now appear dark (Source: F.
Paresce, P. Sartoretti, and
NASA).

The Asteroids

As we continue our journey toward Mars, we now have a difficult stretch in front of us.

Between Jupiter and Mars we will not encounter any large planet, but a band of thousands of small asteroids with diameters of several meters up to 1,000 km. The gravitational forces of Jupiter have prevented these small bodies from forming a single large planet. For ground-based telescopes, the asteroids manifest themselves only as faint points of light. But these tiny planets can be fascinating at close range. Ever since the Galileo probe on its way to Jupiter had close encounters with two of them and took detailed pictures (even finding a moon around the asteroid Ida), interest in these bodies increased significantly. Before its repair, Hubble attempted images of asteroids with larger diameters, where details might have been expected. In 1993 images were taken of the asteroid Fortuna from a distance of 231 million kilometers, but the results were modest. From time to time, asteroids stray close to Earth. On such an occasion in December 1992 Hubble observed the asteroid Toutatis, only a few kilometers in size, at a distance of only 4.4 million kilometers. While the Space Telescope was not able to see more than a point of light, it set a speed record for tracking an object in the sky. Hubble was able to follow Toutatis even though it was traversing one arcsecond every second – a speed that could have easily be seen with a modest ground-based telescope.

The Hubble achieved its greatest success with regard to asteroids in observing Vesta, one of the largest asteroids with a diameter of 525 kilometers, and together with Ceres, Juno, and Pallas one of the best- known. At the end of 1994 the Space Telescope was able to take images during a complete rotation of Vesta, taking 5.3 hours. This object, which would fit comfortably between Baltimore and Boston, was observed from a distance of 252 million kilometers. Under these circumstances, a resolution of about 80 km is quite remarkable, even if it provides only limited scientific insight. But in concert with data from large ground-based observatories, surface maps can be constructed. Now it is known that Vesta is the most geologically diverse of the large asteroids, and the only one with conspicuous bright and dark areas – quite similar to our Moon. Spectroscopy from the ground indicates large regions rich in basalt. Lava must have once flowed on Vesta's surface, meaning the asteroid must have contained a hot, molten interior. One explanation is based on the possibility that radioactive material may have been present during the formation of Vesta, probably left over from a supernova in the vicinity of the location of the soon to be formed sun and planetary system. This radioactive isotope produced energy, melting Vesta's core, and making it a more differentiated body – like the terrestrial planets! In spite of being a bit too small, Vesta could easily be considered the sixth earth-like planet, in addition to Mars, Earth, the Moon, Venus, and Mercury. Shortly after its formation more

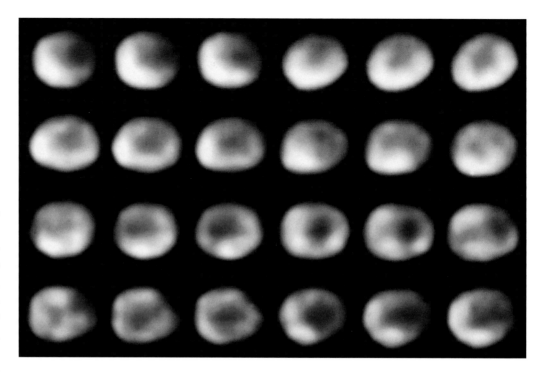

The asteroid Vesta and its rotation. This dwarf has a diameter of only 525 km and is at a distance of 251 million kilometers from Earth and Hubble. Even so, details as small as 80 km across can be distinguished (Source: B. Zellner and NASA).

than 4 billion years ago, molten lava penetrated the surface, cooled again, and has not changed since: In the various shadowy images we see one of the oldest surfaces in the solar system. In contrast, the two asteroids visited by the Galileo probe, Gaspra and Ida, were broken off larger bodies by collisions only several hundred million years ago; they are more indicative of the geological present in the asteroid belt.

Like other bodies of the solar system, Vesta has been hit by smaller asteroids during the last four billion years. The larger collisions broke off some of the crust, exposing deeper layers of olivine.

In one area, a large impact basin can be discerned. Several small pieces from its crust may have made it to Earth, identifiable by the spectral signature of pyroxene, which frequently appears in lava. Quite a number of such small fragments of Vesta are present in the asteroid belt. These cosmic bombs can even be a danger for Earth. The giant planet Jupiter may cause changes in the orbits of small and large asteroids, sending them onto collision courses with Mars or Earth. The picture on the right shows such a fragment of Vesta, which landed in Australia in 1960: the meteorite of Millbillillie, consisting almost completely of pyroxene.

Initiated by the comet crash on Jupiter, the discussion about danger for Earth from these vagabonds in space has intensified. Scientists now search for Near-Earth Objects and attempt to catalog them. But even if a collision with a body large enough to destroy Earth

5 cm
2 in.

is relatively unlikely, Earth like all other planets has to endure continuous impacts of small bodies. These could have catastrophic effects in populated areas.

Mars

With Mars, we have reached the earth-like planets, our immediate neighborhood. Still, this planet is at an average distance of 228 million kilometers from the sun; it has only one tenth of the Earth's mass.

Hubble has delivered the sharpest images of Mars ever taken from Earth and its vicinity. From the many pictures of the red planet taken over Hubble's lifetime, we now know that the climate on Mars has changed significantly since the time of the *Viking* probes in the late 1970s, and that assumptions about the Martian climate based on these data were wrong. Only a systematic survey over longer times can lead to an understanding of its meteorology.

Since the *Viking* probes have fallen silent, the temperature on Mars has plummeted. Globally, it decreased by about 20 kelvin; the planet is now cooler and the atmosphere clearer than before. The reason may be related to the significantly decreased rate of big dust storms. In the first year of the *Viking* visits, two major storms occurred; small dust particles remained in the atmosphere longer than usual. Heated by the sun, these dust particles are the most important source of heat in Mars' atmosphere. The *Viking* probes saw Mars only in this dusty state. Now the atmosphere has become more

transparent and also contains more clouds than in previous years. Water vapor in the atmosphere has crystallized into cirrus clouds, and the planet as a whole has become even cooler and drier than usual. Because the *Viking* probes saw few clouds, they were considered irrelevant, but now they turn out to be important transport mechanisms for water between the north and south pole over the Martian year. The seasonal winds, on the other hand, are responsible for redistribution of dust on the surface of Mars, causing shifts in the shape of the dark regions. After the finer, brighter dust is blown away, the coarser, darker sand remains.

Another argument for a drier atmosphere comes from the increasing ozone excess: Hubble's ultraviolet sensitivity is ideally suited to monitoring the ozone density over the entire planet. It turned out that the ozone excess over the north polar cap (known since the times of the *Mariner 9* probe) had progressed to intermediate and low latitudes. Mars does not have an ozone hole but the opposite; however, the ozone density is still so low that it would provide no protection for terrestrial visitors. Hubble's observations of Mars are important for the planning of future unmanned and especially manned missions to the red planet. First, one would like to land during a season when dust storms are least likely. Furthermore, it is of vital importance to know how cold or warm the atmosphere might be during a landing attempt. If the atmosphere were warmer or colder than expected, it would expand or contract,

A piece of Vesta that fell to Earth in western Australia – after the Moon and Mars (from which meteorites fall to Earth occasionally as well), Vesta is only the third known body from which material is at hand (Source: R. Kempton).

The planet Mars in December 1990 – in the center is Syrtis Major, one of the most prominent dark regions of the red planet. These are not oceans or vegetation, but areas with less sand than brighter areas (Source: P. James and NASA).

causing less or more friction for an entering space vehicle, at worst causing it to burn up or skip back into space.

The three images on the opposite page were taken on February 25, 1995, when Mars was at a distance of 103 million kilometers from Earth and had an angular diameter of 13.5 arcseconds. At Hubble's resolution of about 50 kilometers, several dozen impact craters are visible. Most of the carbon dioxide of the north polar cap has sublimed into gaseous form during the Mars spring, so that only the permanent cap of water ice, several hundred kilometers in diameter, remains visible. In the Tharsis region a crescent-

shaped bright cloud is apparent around the Olympus Mons volcano, which has a diameter of 550 kilometers at its base. Warm afternoon air flowing over this range cools upon descent and forms ice crystals. Similar effects can be seen around three smaller volcanoes in the vicinity. The bottom left part of the center image shows the Valles Marineris, a gigantic system of valleys dwarfing the Grand Canyon; close to the center of this image the Chrysis basin is visible. The third image on the right is dominated by the impressive dark area of Syrtis Major. In general the dust storms of the past have distributed the fine dust predominantly toward the northern hemisphere (top),

Mars 1995: The left image shows an extended desert with high volcanoes, where clouds have formed (particularly on the huge Olympus Mons); the center image shows the vicinity of the huge Mars canyon Valles Marineris; and the right one shows Syrtis Major again (Source: P. James, S. Lee, and NASA).

so that the coarser, darker sand remained mostly in the south, producing structures that can even be discerned with small telescopes.

In principle, our imaginary journey through our solar system would have to end at Mars, since Hubble is almost incapable of observing the two innermost planets, Venus and Mercury. Their orbits around the sun (at distances of about 108 and 58 million kilometers, respectively) lie inside the Earth's orbit, so that they only reach angular distances from the sun of a few tens of degrees, remaining in the so-called zone of avoidance around the sun: Hubble must not look closer than 50 degrees to the sun, otherwise the bright sunlight could burn out its optics and instruments. However, at the beginning of 1995 the opportunity presented itself to observe Venus just inside this 50 degree limit, so the various safeguards were slightly modified and the observations were carried out. The goal was not to take images of Venus's dense carbon dioxide atmosphere – other spacecraft had done that in great detail – but to do spectroscopy. In 1978 the atmospheric probes of the Pioneer Venus Orbiter had found a surprising amount of sulfur dioxide in the atmosphere, which had been interpreted as indicating a major volcanic eruption a short time ago. Since then, the sulfur dioxide concentration has decreased, and Hubble's spectra confirmed this trend. According to these data, there have not been any volcanic eruptions during the last years, which is also confirmed by the radar images of the Venus surface from the *Magellan* spacecraft. The surface did not show any change over the time of the mapping. In addition to the spectroscopic observations, Hubble took several pictures on January 24, 1995, over a distance of 114 million kilometers; these pictures were then combined into a pseudo-color image, showing the familiar patterns in the dense atmosphere of Venus. Darker areas, regions of higher sulfur dioxide concentration, take only four days to circle the planet.

Here we end Hubble's journey through our solar system, with new findings about Venus. Even though the Space Telescope has been established primarily for the observation of very distant objects and galaxies, it has also succeeded in providing important knowledge about our home in the universe.

Venus on January 24, 1995, in the ultraviolet: While polar regions appear bright, lower latitudes show a characteristic pattern of clouds (Source: L. Esposito and NASA).

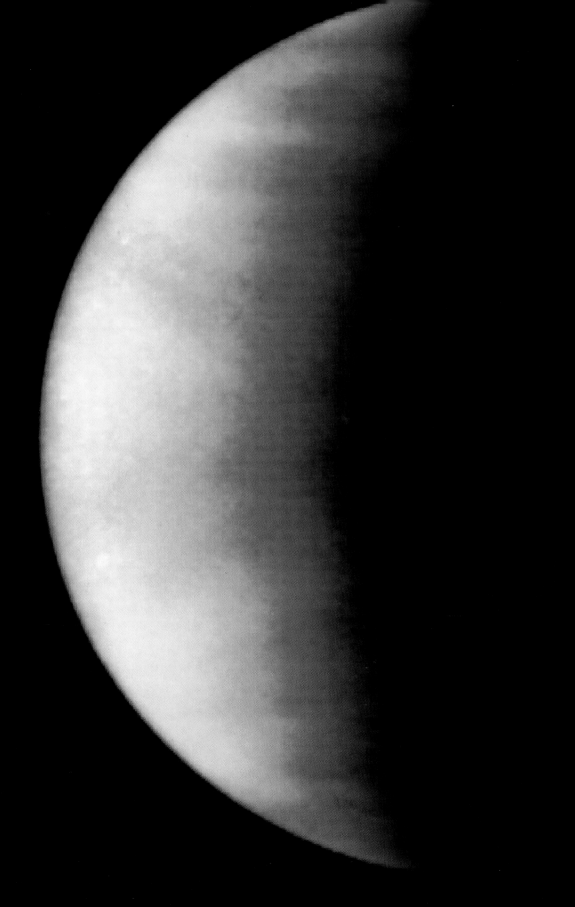

Part 3

Hubble At Work – And Looking Beyond

Working with Hubble

After its shaky start, the Hubble Space Telescope now appears to be a success story after all. During its first five years of operation, about 1000 scientific papers based on its observations have been published in the astronomical literature, in addition to numerous newspaper and magazine articles. Observing time remains heavily oversubscribed, and many more scientists apply than can actually be accommodated.

The efficiency of the telescope, that is the time when images and spectra can actually be taken, has steadily increased from less than 30 percent at the beginning to around 50 percent in 1995, thanks to more experience and improved planning tools. The rest of the time is needed to move the telescope from object to object and to acquire the guide stars for positioning the telescope and the target itself; and part of the time observations remain impossible, as the Earth gets in the way of the telescope during its orbit.

After frequent interruptions during the first years, entries of the telescope into "safe mode" have occurred only three times during the first 17 months after the servicing mission. These safe modes are caused by real or perceived malfunctions. The flight software interrupts normal operations, puts the telescope and its instrument into safe states, and waits for recovery commands from the ground. The latest three occurrences have been caused by small problems with the drive electronics for the solar panels.

Working with the Hubble "has largely become routine," as a very pleased Bob Williams, the sec-

ond director of the Space Telescope Science Institute, notes. "We understand the characteristics and intricacies of the satellite, and we are getting better and better. Soon the planning times will be reduced from a month to about a week." All instruments are working well, and even a faulty part of the Faint-Object Camera, which could not be fixed in 1993, appears to be working again to a large degree.

In addition to the control center at the Goddard Space Flight Center, where the actual "flying" of the satellite takes place, the Space Telescope Science Institute (ST ScI) serves as the center for scientific operations and as the interface to the astronomical community. It is located on the Homewood Campus of Johns Hopkins University in Baltimore, Maryland, and it is operated by the Association of Universities for Research in Astronomy (AURA) under contract with NASA. Currently, about 400 astronomers, scientists, engineers, software analysts, and support personnel work there, including a majority of personnel from AURA, but also from the Computer Sciences Corporation (CSC) and the European Space Agency (ESA).

Let us follow the course of an observation with the Hubble from its inception to completion, and we will begin to appreciate the complexity and difficulty of the various tasks required to carry out a single project. Every year, ST ScI issues a Call for Proposals to the astronomical community. Scientists from all over the world can propose observations of their favorite objects with the Hubble. About a thousand proposals,

requesting over five times as much observing time as is actually available, are usually received. Once all proposals have been collected, the institute organizes panels of independent scientists to rank the proposals according to scientific merit. Given the available observing time, about 400 proposals are typically accepted for execution, while the rest have to be rejected. Contact scientists and program coordinators are assigned to each proposal; they will advise the proposers on the details of the next steps, provide support, and communicate with them on all matters relating to their proposal.

Once a proposal has been accepted, the details of the observations have to be worked out between the proposer and the institute staff, ranging from the precise target coordinates to instruments, filters, and exposure times. When these details have been defined, the actual planning process begins. Taking into account the positions of the targets in the sky and all other observing parameters of all available proposals, a detailed calendar of events is established for a full week of observing. This may sound simple, but it is a rather complex process, as it attempts to optimize observing time as much as possible. This schedule is then translated into individual spacecraft commands and eventually loaded into the Hubble's on-board computers and executed. The presence of the investigator during the actual observation is normally not required; only for time-critical observations, or where complex maneuvers are needed, can they be of help. Through a circuitous route involving two communications satellites, the raw data arrive back at the Space Telescope Science Institute – the observation has been done.

The data flow from Hubble (left) to the Space Telescope Science Institute (right): The first intermediary is a Tracking and Data Relay Satellite (TDRS) in geostationary orbit, sending the data to a ground station in New Mexico. From there, they travel to another communications satellite, and then to the Goddard Space Flight Center in Greenbelt, Maryland. Finally, they arrive via cable at the institute in Baltimore, Maryland (Source: ST ScI).

The data from Hubble arrive here: at the Space Telescope Science Institute (left) on the Homewood Campus of Johns Hopkins University. The control room is shown on the right (Source: D. Fischer and ST ScI).

But the work is far from over. The observing data arrive in a form that would make them extremely difficult to use. First they have to be extracted from the raw telemetry, which also contains lots of engineering data necessary to run the telescope, but only of limited interest to the observer. Now the science data are subjected to automatic processing, which converts the numbers produced by Hubble's instruments into physically meaningful quantities; the data are calibrated. To do this, ST ScI maintains not only an elaborate software system to analyze the data, but specific calibration observations for each instrument are planned, carried out, and analyzed in detail in regular intervals. After a quality check of the data by one of the institute scientists, confirming that the observations were carried out correctly, the data are ready to be sent to the observer, usually within a few days after the observation. They also become part of the Hubble Data Archive, maintained at ST ScI.

Normally, the Hubble data "belong" to the proposer of the respective observations for a proprietary period of one year; during that year, though stored in the Hubble Data Archive, they are not available to anybody else. The investigator is expected to analyze the data in depth and hopefully reach new scientific insights during that time, and to publish the results in the astronomical literature. During this analysis phase, the observer may arrange visits to ST ScI for expert help and advice. Once the proprietary period is over, all data become public and can be used by anyone. At this time, the individual project can be considered complete. From preparation of the first proposal through the publication of the results, it may well have taken two years or even longer.

The Hubble Data Archive at ST ScI, where all telescope data end up, is not meant to be a mausoleum. Great care has been taken to make it a dynamic and valuable resource, accessible to the astronomical community as well as the general public. Data arrive

A sophisticated but very user-friendly software interface was developed on top of the actual archive system, to enable astronomers and the public to access Hubble data electronically over the Internet. Not only the data themselves but all supporting information is available on-line. Since the opening of the archive, retrieval rates have grown to tens of gigabytes per day. Astronomers use the archive either by "browsing" to see what observations have been made, and what may be available, or they can actually retrieve significant amounts of data for their own research projects; these larger projects are usually part of so-called Archival Proposals, which are supported by Space Telescope Science Institute in a manner similar to Hubble observing proposals.

But ST ScI is not the only place providing support to the astronomical community for observations with the Hubble. In Europe, the Space Telescope European Coordinating Facility (ST-ECF) shares space with the headquarters of the European Southern Observatory

Where the data end up: the archives in Baltimore, Maryland, and Garching, Germany. Data are stored on large optical disks, holding six gigabytes each. In Baltimore, they are mounted in robotic juke boxes (right). In Garching (left), where demand is lower, requested disks are mounted in individual drives by an operator (Source: D. Fischer).

at the archive at a rate of about one gigabyte (one billion bytes) per day, and once new instruments with larger and faster detectors are installed on the Hubble during the next servicing mission, this rate will increase severalfold. The data are stored on optical disks with a capacity of six gigabytes each; the individual platters are mounted in large robotic juke boxes, so that direct access to all data is available with relatively short access times. At the end of the projected lifetime of the Hubble, the archive will contain several tens of terabytes, or trillions of bytes!

(ESO) in Garching near Munich, Germany, and is operated jointly by ESO and the European Space Agency (ESA). This institution assists European proposers in preparing their activities with the Hubble, and in analyzing their data. In addition, a copy of the Hubble Data Archive is maintained there.

In our discussion of the archive of Hubble data, we have seen that these data are not only available to professional astronomers. Other activities reach beyond this community to the general public as well.

Early on, Riccardo Giacconi, then director of the Space Telescope Science Institute, decided to make a small amount of Hubble's observing time available to amateur astronomers. Proposals from amateurs were solicited and selected in a similar fashion to those from professionals.

The programmer George Lewycky from New York, for instance, had the opportunity to experience the adventure of using Hubble first hand. His project to obtain spectra of Saturn's moon Titan was approved in 1992. Results of *Voyager I* had fascinated him and provided the basis for his proposal. His observations were taken on September 9, 1993. Lewycky experienced all the stages of work with Hubble: the joy and excitement of receiving the data, the difficulties of analyzing the complex spectra, and the fascinating discussions with astronomers working on similar questions. The amateur project on the Hubble had generated enthusiastic ambassadors to the public.

Going even further, NASA in general and the Space Telescope Science Institute in particular have undertaken a broad project of outreach, trying to convey the excitement of the Hubble project to the general public. Activities include extensive contacts with teachers and schools, public talks, press releases, printed materials, TV appearances, and planetarium shows, all centered on the Hubble Space Telescope and the fascinating results obtained with its help.

Hubble's Next Decade

Hubble's design lifetime was defined as fifteen years. A third of that time span had been reached by 1995. During the remaining years, at least three more servicing missions are planned. Aside from routine replacements of degraded or failed components, as was also done during the 1993 mission, emphasis will be put on the replacement of scientific instruments for newer, more modern developments, and on raising Hubble's orbit when necessary.

In this age of budgetary problems, the Hubble project has been asked to share the burden as well. Shortly after the servicing mission of 1993, almost as a reward for the successful repair of the Space Telescope, fairly drastic budget cuts were implemented, so that, for instance, work on certain replacement parts had to be curtailed – parts that were supposed to guarantee Hubble's fifteen-year lifetime. This may endanger its continued operation, according to the Space Telescope Users' Committee. In light of the past successes and the great public interest in results of the repaired telescope, these cutbacks are difficult to understand. The Space Telescope Users' Committee thinks there was an erroneous assumption, both inside and outside NASA, that the budget could be arbitrarily cut without putting the current scientific operations in danger – in reality, the committee did not believe that any more jobs could be cut.

Of course there are areas where modernization and increased efficiency can lead to economies without jeopardizing ongoing operations. After all, the concept for daily operations of the telescope goes back to the 1970s and early 1980s. On a larger scale, the basic philosophy of satellite operations is being examined. NASA headquarters, the Goddard Space Flight Center, and the Space Telescope Science Institute have begun a project called Vision 2000 that will lead to substantial streamlining of operations, elimination of duplication, and application of state-of-the-art techniques. This may benefit not only the Hubble project but other satellite operations as well. On a smaller scale, old computer systems, which are expensive to maintain, have been replaced with smaller, cheaper, newer, and more capable ones, and large parts of the ground system software have been rewritten, to be more user friendly and more functional.

The second servicing mission, planned for February 1997, will have the main goal of exchanging the two existing spectrographs with the Space Telescope Imaging Spectrograph (STIS) and the Near-Infrared Camera and Multi-Object Spectrograph (NICMOS). Since only four spacewalks are planned for this mission, it may be a bit simpler for the astronauts than the first servicing mission. However, the testing and commissioning of the new instruments will be much more complex, while budgets and time available are lower. In addition to the two new instruments, one of the three Fine Guidance Sensors is planned to be exchanged for a new but essentially identical one, as the old ones have started to show degrees of wear that affect the telescope's ability to point accurately.

The New Instruments STIS and NIC

STIS stands for Space Telescope Imaging Spectrograph, a completely new instrument which will completely replace the two current spectrographs, and provide more extended wavelength coverage, from 105 nm through 1100 nm. STIS is a two-dimensional spectrograph: it can simultaneously take spectra of all objects or areas positioned on its long slit – up to a thousand, providing enormous time savings compared to spectra from single points in the sky, to which the Hubble was confined up to now. Its sensitivity also exceeds that of the old instruments, by more than a factor of 10 in the far red, and of 2 in the ultraviolet. In addition, STIS contains its own camera, which will be used primarily for target acquisition, but which can also be used on its own.

The truly new camera for Hubble will be NIC, the Near-Infrared Camera. For the first time, Hubble will be able to extend its range deeper into the infrared region of the spectrum, and to apply its acuity and vantage point above the Earth's atmosphere to images and spectra up to a wavelength of 2.5 μm.

At these wavelengths, which are about five times longer than visible light, the sky looks quite different and extremely fascinating. Extremely distant galaxies become visible, dense dust clouds, where new stars are forming, are now transparent, while certain areas on bright planets like Jupiter appear darker. This technology has been available for about a decade, and can now be used to take images of all these wonders with CCD-like cameras. The chips developed for NIC have also been a hit for ground-based observatories – but stationed in space they have an enormous advantage: Hubble flies above the airglow, and therefore experiences an infrared sky background which is 100 to 1000 times darker than for terrestrial telescopes. With NIC, Hubble can undertake a deep survey in search of the progenitors of today's galaxies in the early universe, their light being strongly shifted into the red by the expansion of the universe.

NIC consists of three independent cameras and three spectrometers for the various wavelength ranges. In total, three chips are required; they are identical mercury-cadmium-telluride detectors with 256×256 pixels. These units can be used separately or in combination, and have their own microprocessors.

STIS and NIC are planned to be installed on the Hubble during the second servicing mission in 1997.

Another important task during the second servicing mission will be the exchange of at least one of the aging tape recorders, used to buffer science and engineering data before their transmission to the ground, for solid-state recorders providing a tenfold storage capacity. This makes real-time links between the telescope and the ground less critical, as more data can be stored on board, improving scientific productivity. This is particularly important for the new instruments, which will produce significantly more data than the older instruments. Furthermore, two gyroscopes are scheduled for routine replacement, which was already done with different sets during the 1993 mission.

In preparation for boosting Hubble's orbit during a later mission, a test may be conducted in 1997. This test would entail gentle test firings of the shuttle's thrusters with Hubble attached to the cargo bay and the solar panels extended. The reason for this test lies in the risk of stowing the solar panels and then redeploying them on the occasion of an orbit boost. If the test is successful and confirms the structural models for the panels, the real boost may be executed during the third servicing mission. At that time, the solar activity will be near a maximum again, Earth's atmosphere will be extended, and the friction on Hubble will be greater than average, so that the boost will be required.

The third servicing mission is planned for 1999. During this mission, the Faint-Object Camera is scheduled to be replaced by the so-called Advanced Camera. The replacement of the Faint Object Camera also has a political component, since it represents part of the investment in the telescope by the European Space Agency (ESA). The current memorandum of understanding between NASA and ESA will run out in 2001. However, negotiations are under way

The Advanced Camera for Surveys

The Advanced Camera for Surveys, or ACS, actually contains three cameras. The "wide-angle" channel features a CCD chip with 16 million pixels – 25 times as many as a single chip of the WFPC-2 – and a field of view of 3.3 × 3.3 arcminutes; it is mainly sensitive for red light. Two channels for high resolution have chips with one million pixels each, with each pixel covering an area of only 0.03 arcseconds projected onto the sky; therefore, they take full advantage of Hubble's resolution. One of the channels ranges from 200 nm to 1 μm, the other from 115 to 170 nm in the far ultraviolet. All three cameras will be 4 to 8 times as sensitive as the WFPC-2 and the FOC, enabling them to see *all* light-emitting galaxies along the line of sight up to the edge of the universe – as long as budgetary problems do not cause a reduction of its planned capabilities. Like STIS and NIC, the Advanced Camera will correct the spherical aberration of Hubble's main mirror internally, so that COSTAR becomes unnecessary.

to extend the fruitful cooperation until the end of Hubble's mission. As part of these negotiations, a new advanced instrument for the servicing mission in 2002 could be provided by ESA alone, or in cooperation between ESA and NASA. While the American side is highly interested in a new instrument from the Europeans (who have already defined two futuristic new concepts for it), there are also plans to install yet another American instrument in 2002 which would be complementary to the ESA contribution. A major operation on the satellite itself is also planned for the 2002 mission: the present flexible solar arrays are to be replaced by smaller, stiffer and more powerful ones, developed for the commercial communication satellite program Iridium. More important are the upgrades to its scientific payload. Only the WFPC2 will remain on board; all of the other instruments will be replaced by 1990s technology. It will make no sense, therefore, to turn Hubble off when its nominal lifetime ends in 2005. Instead, the final servicing mission in 2002 is now intended not only to complete the upgrade of scientific hardware but also to prepare the satellite for a long final phase of operations, without any further visits. This will give NASA and the world astronomical community additional time to conceive and build a successor.

Since it appears unlikely that a successor to the Hubble will be ready by the end of the nominal fifteen-year lifetime in 2005, there have been cautious plans to extend the Hubble mission. Such a scenario could entail a final servicing mission in 2005, which would be used to boost the orbit for one last time and change a few minor parts at most; no new instruments would be foreseen. From that time onward, activities would be limited to normal scientific operations, no new developments, and only the most critical maintenance, to minimize overall cost. Such a runout period could last until about 2012.

What Comes After Hubble?

It is still unclear what will happen to the Hubble Space Telescope after its mission has been declared complete, assuming it continues to work well and does not undergo a catastrophic failure, for instance being hit by a tiny meteorite. It appears likely that at the end of the mission, sometime between 2005 and 2012, the Hubble would be retrieved from orbit and brought back to the ground.

Scientifically successful and worthwhile as it already has been, and hopefully will continue to be, Hubble is an extremely large and expensive project. Every servicing mission, for instance, costs a few hundred million dollars. Therefore, some have called for smaller, simpler, but more frequent missions. This would also allow shorter lead times, and the incorporation of new technologies.

Of the four "Great Observatories" NASA planned, two are flying, the Gamma Ray Observatory (GRO) and the Hubble Space Telescope. AXAF, the X-ray satellite project, has undergone substantial cutbacks, and consequently the project has been shrinking, as has the satellite; a launch is expected in a few years. The Space Infra-Red Telescope Facility (SIRTF) is still fighting for funding. Aside from a number of small projects, no plans exist for a telescope in space, operating in the visible and immediately neighboring parts of the spectrum. There are no concrete plans for a successor to the Hubble.

From the scientific point of view, it will be incumbent on the astronomical community to define the priorities and goals for future space missions.

Should the successor to the Space Telescope be a "Super-Hubble" with a larger mirror, perhaps six to ten meters in diameter? Or should there be a number of smaller missions dedicated to answering individual specialized questions? Should it be a "normal" telescope with a filled aperture, or should it be an interferometer, which would allow increased angular resolution, to probe details of the structure of distant objects and to determine positions and movements of individual stars to very high accuracy, but at some sacrifice of its ability to reach down to very faint objects? What wavelength range would provide the most scientific information for the funds available? Obviously, these questions have to be answered, and soon. In addition, it will be necessary to support these decisions unanimously and forcefully, in order to achieve public support and funds, as the history of the Hubble shows as well. For many, the prospects after the end of the Hubble project appear bleak. The danger of being left without a successor, without a modern tool to continue research in the optical and near-optical wavelength ranges, seems very real. After all, any medium- or large-scale project in space would require lead times of one or two decades. It is high time to start the planning for space astronomy after Hubble.

In light of this situation, planning for an extended life for Hubble appears to be a necessity. Bob Williams, the director of the Space Telescope Science Institute, is therefore in favor of "extending the lifetime of Hubble, as long as it is scientifically

Robert Williams, the second director of the Space Telescope Science Institute, who replaced Riccardo Giacconi in 1993 (Source: ST ScI).

productive, which is very possible."

NASA has also started the process of defining a future astronomy mission, to succeed the Hubble Space Telescope. Several committees have begun to weigh the various possibilities and approaches for "HST and Beyond," as one of these committees is called. The common theme of "origins" has emerged from these deliberations: origin of the universe, origins of galaxies and stars, origin of life (planets around other stars). But time is short if we wish to assure a seamless transition from the Hubble project to its successor.

It appears that the answer to the question "What Comes After Hubble?" cannot be given yet. We hope that optical astronomy will not be confined again to the surface of the Earth by the beginning of the twenty-first century, in spite of the advent of larger and more powerful telescopes there, as the capabilities of space-based telescopes are unique in their resolving power and wavelength coverage. We hope that after the end of the Hubble project we will not be deprived of the exciting images and data from so many interesting celestial objects. We hope that the fascinating images from the Hubble Space Telescope, a small fraction of which can be seen in this book, will not remain a singular event in the history of astronomy.

Part 4

Appendix

Want to See More?

The Space Telescope Science Institute routinely makes Hubble Space Telescope images, press releases and other information available electronically. You can access GIF and JPEG images, captions and press release texts via the World Wide Web at URL http://www.stsci.edu/pubinfo/PR/95/44.html, or via links in http://www.stsci.edu/Latest.html and http://www.stsci.edu/pubinfo/Pictures.html.

High-resolution (300 dpi JPEG) versions of many images are also available.

You can get Space Telescope Science Institute press release texts and other information sent to you automatically be sending an e-mail message via Internet to listserv@stsci.edu. In the body of the message (not the subject line) type the words subscribe pio [your name]. Don't use user/account names: if your name is John Smith, type subscribe pio John Smith. The system will reply via e-mail with a confirmation of your subscription.

Further Reading

History of Astronomy and General Introductions

Jean Adouze and Guy Israel, eds.: The Cambridge Atlas of Astronomy, Third Edition. Cambridge: Cambridge University Press, 1994.

William K. Hartmann: The Cosmic Voyage: Through Time and Space. Belmont, Calif., Wadsworth, 1992.

William J. Kaufmann: Universe, Fourth Edition. New York, W.H. Freeman, 1993.

David Leverington: A History of Astronomy: From 1890 to the Present. London, Springer-Verlag, 1995.

John Norris: The Norton History of Astronomy and Cosmology. New York, W.W. Norton, 1994.

Richard Preston. First Light. Boston, Atlantic Monthly Press, 1987.

Archie Roy, ed.: Oxford Illustrated Encyclopedia of the Universe. Oxford, Oxford University Press, 1992.

Hugh Thurston: Early Astronomy. New York, Springer-Verlag, 1994.

The Hubble Space Telescope

Eric Chaisson: The Hubble Wars: Astrophysics Meets Astropolitics in the Two-Billion-Dollar Struggle over the Hubble Space Telescope. New York, HarperCollins, 1994.

Carolyn Collins Petersen and John C. Brandt: Hubble Vision: Astronomy with the Hubble Space Telescope. Cambridge, Cambridge University Press, 1995.

Robert Smith: The Space Telescope: A Study of NASA, Science, Technology and Politics. Cambridge, Cambridge University Press, 1989.

Edwin Hubble

Gale Christianson: Edwin Hubble: Mariner of the Nebulae. New York, Farrar, Straus & Giroux, 1995.

Cosmology

John Barrow: The Origin of the Universe. New York, Basic Books, 1994.

Albert Einstein: The Meaning of Relativity, Fifth Edition. Princeton, Princeton University Press, 1956.

Stephen Hawking: A Brief History of Time. New York, Bantam, 1988.

Rudolph Kippenhahn: Light from the Depths of Time. New York, Springer-Verlag, 1987.

Jean-Pierre Luminet: Black Holes. Cambridge, Cambridge University Press, 1992.

P.J.E. Peebles: Principles of Physical Cosmology. Princeton, Princeton University Press, 1993.

Michael Rowan-Robinson: Ripples in the Cosmos: A View Behind the Scenes of the New Cosmology. New York, W.H. Freeman, 1993.

Joseph Silk: The Big Bang, Second Edition. New York, W.H. Freeman, 1988.

Goerge Smoot and Keay Davidson: Wrinkles in Time. New York, Morrow, 1994.

Kip S. Thorne: Black Holes and Time Warps: Einstein's Outrageous Legacy. New York, W.W. Norton, 1994.

Trinh Xuan Thuan: The Secret Melody. New York, Oxford University Press, 1995.

Robert Wald: Space, Time and Gravity: The Theory of the Big Bang and Black Holes, Second Edition. Chicago, University of Chicago Press, 1992.

Steven Weinberg: The First Three Minutes. New York, Basic Books, 1977.

Stars and Galaxies

Erika Bohm-Vitense: Introduction to Stellar Astrophysics. 3 volumes. Cambridge, Cambridge University Press, 1989.

Alan Dressler: Voyage to the Great Attractor: Exploring Intergalactic Space. New York, Knopf, 1994.

David Eicher, ed.: Deep-Sky Observing with Small Telescopes. New York, Enslow, 1990.

Nigel Henbest and Heather Couper: Guide to the Galaxy. New York, Macmillan, 1994.

Edwin Hubble: The Realm of the Nebulae. New Haven, Yale University Press, 1985.

Donald E. Osterbrock, ed.: Stars and Galaxies: Citizens of the Universe. Readings from Scientific American. New York, W.H. Freeman, 1990.

Jean-Claude Pecker: The Future of the Sun. New York, McGraw-Hill, 1992.

Gareth Wynn-Williams: The Fullness of Space: Nebulae, Stardust and the Interstellar Medium. Cambridge, Cambridge University Press, 1991.

The Solar System

J. Kelly Beatty and Andrew Chaikin, eds.: The New Solar System, Third Edition. Cambridge, Mass., and Cambridge, England, Sky Publishing and Cambridge University Press, 1990.

Ronald Greeley: Planetary Landscapes, Second Edition. London, Chapman and Hall, 1993.

William K. Hartmann: Moons and Planets, Third Edition. Belmont, Calif., Wadsworth, 1992.

John H. Rogers: The Giant Planet Jupiter. Cambridge, Cambridge University Press, 1994.

Carl Sagan: Pale Blue Dot: A Vision of the Human Future in Space. New York, Random House, 1994.

Index

30 Doradus, 94, 95

A

aberration, 40, 44
aberration, chromatic, 13
aberration, spherical, 39, 41, 42, 45
Advanced Camera for Surveys (ACS), 162
airglow, 24
Akers, Tom, 53
Allen Commission, 44
amateur astronomers, 159
Andromeda galaxy (M31), 91
Anglo-Australian Observatory, 18
Arp 220, 82
asteroids, 125, 145, 146

B

Bahcall, John, 53
Barnard, Edward E., 15
barred spirals, 63
Beta Pictoris, 51
black holes, 85–87, 89, 93, 120
blue giants, 120
brown dwarfs, 95, 96

C

calendar systems, 13
Canada-France-Hawaii Telescope, 18
Capella, 51
Cartwheel Galaxy, 77, 78
Cepheids, 16, 67, 68
Ceres, 145
Cerro Tololo Interamerican Observatory, 18
Charge-Coupled Devices, 34
Charon, 49, 128, 129
Cirrus nebula, 115, 121
collision of galaxies, 78, 80
Coma cluster (of galaxies), 69
Comet P/Shoemaker-Levy 9, 137, 139
control software, 43
Corrective Optics Space Telescope Axial Replacement (COSTAR), 45, 46, 51, 53, 55, 60, 89
cosmological constant, 64–66, 70
cosmology, 9
Crab nebula, 121
cross staff, 13

D

dark matter, 66, 85, 101

E

density of the universe, 65, 66
Deuterium, 51
Doppler effect, 63

E

Einstein Cross, 49, 84
ESRO, 21
Eta Carinae, 117, 123
Europa (moon of Jupiter), 143
European Southern Observatory (ESO), 18, 158, 159
European Space Agency (ESA), 29, 30, 155, 162, 163
expansion of the universe, 17, 64, 65, 70

F

Faint Object Camera (FOC), 28, 29, 36, 37, 39, 72, 89, 162
Fine Guidance Sensors (FGS), 27, 34, 39, 43, 125
first light, 35, 36, 39
flare, 96
Fortuna (asteroid), 145
Fraunhofer, Joseph von, 14
Freedman, Wendy, 68
Friedmann, Alexander, 17